2030: THE COMING TUMULT

2030: The Coming Tumult

Unlimited Growth on a Finite Planet

Richard M. Mosey

Algora Publishing
New York

Library of Congress Cataloging-in-Publication Data —

Mosey, Richard M., 1950-
    2030, the coming tumult: unlimited growth on a finite planet / Richard M. Mosey.
        p. cm.
    Includes bibliographical references and index.
    ISBN 978-0-87586-744-1 (trade paper: alk. paper) — ISBN 978-0-87586-745-8 (hard
cover: alk. paper) 1. Global environmental change. 2. Natural resources—Management. 3.
Environmental degradation. 4. Overpopulation. 5. Conservation of natural resources. I.
Title.
    GE149.M69 2009
    363.7—dc22
                            2009032345

Printed in the United States

To Jean, for everything

TABLE OF CONTENTS

PREFACE

This book tells the story of the current climate change crisis, how it evolved, and why we shouldn't be surprised when little or no action is taken.

This volume focuses largely on water. According to the 1969 hit song by the 5th Dimension, the "Age of Aquarius" was to be an era of harmony and trust and understanding. It appears to be evolving instead into the "Age of Nefarious," an age of conflict and theft and genocide due in large measure to careless, short-sighted or unapologetically selfish abuses of finite life-giving resources as well as global over-reliance on fossil fuels. The geological, technological, demographic and political aspects of the looming shortfalls are all discussed in this book.

A second volume will cover the influence of monotheistic religions, culture, and civilization on the climate debate and the secular religion of "growth" and its influence on how we approach living on a finite planet. It will also examine issues surrounding the globalization debate, the cult of consumption in the United States, our country's infrastructure, a rise in regional and global conflicts over resource allocation, and what kind of world we are likely to leave to our children and grandchildren as well as their growing estrangement with the natural world.. Taken together, these volumes aim to provide a comprehensive view of the substantial obstacles facing those who wish to avoid the inevitable consequences of a rapidly changing planet.

> He walked out in the gray light and stood and he saw for a brief moment the absolute truth of the world. The cold relentless circling of the intestate earth. Darkness implacable. The blind dogs of the sun in their running. The crushing black vacuum of the universe. And somewhere two hunted animals trembling like ground foxes in their cover. Borrowed time and borrowed world and borrowed eyes with which to sorrow it.
> — *The Road*, Cormac McCarthy

INTRODUCTION

"Prediction is very difficult, especially about the future." — Niels Bohr

Despite the admonitions of Mr. Bohr, the *Coming Tumult* books intend to do just that — predict what our world will look like in the year 2030. In historical time, of course, this span is a blip at best. In geologic time, it's less than a blip. I will be 80 years old and, most probably, dead. The college kids who file past my window each morning toting their backpacks and obligatory plastic water bottles will be in their early to mid-forties — the prime of life but participating in a world much different from what we know today.

When I began writing this book, George W. Bush was president. The economy was generally bullish — or at least that was what the White House wanted us to believe. Today, Barack Obama is president. The Bush economy has tanked and the reverberation has been felt around the world. The American-style national and global free marketeering has already revealed itself for what it is: a shadowy world of enormous profits for the wealthy and a disaster for average Americans and other national economies seduced by globalization. The coming catastrophe was apparent for those who could read the signs and delve beyond the spin. The rapidity, though not the end result, was breathtaking. The true extent of the economic fallout has not yet abated. The costs have been and will be enormous. Leaders of other world economies have noted America's role in this meltdown and will not soon forget it.

Climate change as an issue has all but dropped off the political radar as more immediate economic woes have come to the foreground. President Obama has conceded that the national deficit will skyrocket into the *trillions*, figures so enormous they are difficult for human beings to comprehend.[1] Effectively fighting climate change will cost *trillions* more.[2] Americans will be unwilling and unable to supply the funds needed to make the necessary changes. In the perverse political atmosphere in which we live, many conservatives are fairly crowing that any substantial action on climate change is dead because rescuing the tattered economy that conservatives themselves helped to create will suck up all available funds. And they're probably right. Future problems, no matter how dire, do not exist as long as short-term financial woes take center stage.

At the outset, I would like to state that I have never been prone to conspiracy theories or extreme levels of paranoia. I write this book reluctantly. My original intent as a writer was exactly the opposite — to supply humorous stories about human foibles gleaned from newspapers around the world. I am pleased to report that the human comedy goes on uninterrupted, and that is an endeavor to which I planned to return after completing this book. But I became distracted. My interest in climate change, population control, water scarcity, and other problems was probably higher than average, but I'd be the first to admit that I am no expert. Like most Americans, my sources of information have typically been the popular press, television, and the Internet. But in my international search for humor, I couldn't help but notice the increase of more disturbing events chronicled in the global press, from China to Argentina. The transformation — which will be extreme and will involve many actors — has already begun. Unlike in the cowboy movies of my youth, sometimes the white hats are indistinguishable from the black. The shades are gray and many *eminences grises* lurk in the corners. Generally, the American public is too uninformed and too preoccupied with its own problems to grasp the forces at work whose actions will dramatically alter the world as we know it, precipitating a plunge into chaos at best, if not actual extinction.

The problems are both simple and complex. I will delve into the complexity in the chapters that follow. But the premise is simple: The Earth has reached or passed its carrying capacity in terms of both industrialization and population growth. The planet's ecosystem will not be able to function properly beyond 2030 to support the inevitable population growth, especially in the fast-growing Middle East and Africa. Globalization, driven by American-style free market capitalism which has reached the status of a secular religion, cannot survive without continuous short-term global economic growth that will degrade the Earth's environment beyond its capability to bounce back. While mainstream economists

---

1 Don A. Rich, Ludwig von Mises Institute, "The Real Cost of a Full Bailout," Aug. 22, 2008.

2 The true cost of fighting climate change is, of course, unknown , according to the Congressional Budget Office. Some economists have estimated it would cost the U.S. government at least 1 percent per year of the Gross Domestic Product. The estimated GDP in the U.S. in 2009 was estimated at $14,075 billion. The U.S. generates 30 percent of the world's total GDP.

and corporate elites hurl the world farther down this path, they disregard the political, cultural, and ecological repercussions of their errant policies — partly infused by an overly optimistic George Babbitt "can-do" recklessness and, less naively, in pursuit of profiteering that will allow them to survive in what the more pensive among them have realized will be an increasingly privatized and Darwinian — even Hobbesian — world. The year 2030 will not be a good time (if ever there was one) to be poor — and declining standards of living will be the norm for at least 80 percent of the American people.

Sadly, the process has already begun. The United States in the not-too-distant future will become indistinguishable from Peru or Ecuador — the rich (increasingly the top 10 percent of the U.S. population) and the very rich (about .01 percent of the country) will hunker down behind walls patrolled by private security forces while the less advantaged majority try to survive as best they can. Even former Federal Reserve Chairman Alan Greenspan, a life-long libertarian, has expressed concern about the growing income inequality that has continued to escalate, with a few slight exceptions, since the Ronald Reagan Administration and reached it apogee in the George W. Bush Administration. Capitalism cannot survive without growth. The Earth cannot handle that growth. It's capitalism versus nature in a battle to the death. Corporate elites, while making puffy public relations noises about environmental concerns with sticky-sweet television advertisements, have no intention of cutting either economic growth or environmental pollution — an assertion that this book will illustrate.

Optimism is a good thing until it kills you. The relentless refusal to face facts has put everyone at risk: the world's poor, the ever-shrinking and increasingly desperate American middle class, as well as the delicate and beautiful planet that has, at least until recently, provided humanity with an awesomely complex and efficient system which nurtured our species to unparalleled heights *free of charge* but which we are now willfully destroying like a fine Swiss watch pounded by a hammer. So who wins? Not me and probably not you — unless you are part of the transnational clique who gathers annually in Davos, Switzerland. These scions of privilege are squirreling away their financial acorns to survive the bleak ecological and economic winter that they know is coming. This is, of course, to their own advantage and assures the prosperity of their progeny. They are as smart as they are myopic and greedy. Their children, it seems, are worth more than your children. The American people have been very useful idiots, supporting the dismantling of their own economy in direct conflict with their own interests as economic and the more emotionally-charged social and religious issues were made to intersect. These corporate and political elites, who until recently we had elevated to rock star status, knew which buttons to push to validate their self-serving view of the world. Like lemmings, we have followed them over the cliff.

In researching this book, I have made every effort to review all sources with an open mind: I am more interested in the truth than a screed. I have reviewed the pro-globalization and anti-globalization literature; books promoting and decrying our consumer culture; pro-religious and anti-religious points of view. All will be discussed in detail in this three-volume series.

There have been books *ad nauseam* about the perils of global warming as well as the benefits of free-market capitalism and its spawn, globalization. The science is clear but scientists, many of whom are screaming out of frustration about the ecological problems, have underestimated the power of the political filter. Neither Republicans nor Democrats will publicly discuss income inequality from the top down, even though it is the most important national economic issue of the 21st century. During the recent economic expansion, the elite became substantially richer while the middle class went on a downward slide to greet their new brethren, the poor.

We will become a third-world America as soon as 2030, after it is too late for Joe Public to repel the juggernaut. Scientists, as a group, are not saints, and saintliness is not encouraged in these free-marketeering times. Some have become shills for global corporations; they sell their scientific souls for mammon. No surprise there, and a similar pattern is shown by the many physicians on the payroll of large pharmaceutical companies (an industry which has consistently been the most profitable sector of the American economy). Universities, once a bastion of rational intellectual thought, have also been seduced — sometimes out of necessity in this privatized world — into becoming subsidiaries of the corporations that can provide needed capital. We live in an age drowning in information but bereft of wisdom and an understanding of the ultimate consequences of our actions.

Scientists never had a chance. Their standards are set too high, making any research vulnerable to the political spin generated by the ever-increasing corporately-oriented "think tanks" with laudable-sounding names but financed by a plethora of energy interests who demand *100 percent definite proof* about their scientific research — the actual fire rather than the voluminous smoke. Climate is complicated. Ask any meteorologist who tries to extend a forecast beyond three days. Spin doctors realized this weakness long ago (think "Big Tobacco") while carefully planting seeds of doubt to confuse the gullible public. The ploy, as expected, has been magnificently successful so that the majority of Americans now question the validity of the science even though all the canaries in all the coal mines are gasping for air or dropping dead. Journalists, under the guise of fairness, use the tried and true "he said/ she said" format which tends to give equal weight to both sides — and significantly skews the truth.

Science has other limitations. Most scientists don't talk to one another to get a clearer picture of the systemic whole. Science has become micro as scholars concentrate on ever smaller disciplines — from, say, crabs in general to a certain genus of crab to an individual species. While knowledge has increased, the interconnectedness remains understudied and generally unexplained to the general public. It's tough, even for scientists, to keep up. Scientists also know who butters their bread: when in doubt, get a grant and do more studies — until it's too late to do anything meaningful. Unfortunately, the coming ecological disaster was recognized long ago — at the time when action would have been relevant. While it's horrible to contemplate, many scientists, scholars, and writers have already realized that the window of opportunity is rapidly closing. Spin has won the day. You can park the car, recycle your waste, and try to be a responsible global citizen, and it won't amount to a spit in the ocean. But that's not sufficiently optimistic, not Babbittism, and somehow vaguely un-American. Something will save us. Technology will save us (wrong) and, if not, God will save us (also wrong). In fact, both technology and the fundamentally religious conspired to create many of the problems in the first place — as will be discussed in the second volume.

Perhaps the greatest and most succinct American philosopher of our age was newspaper editor H. L. Mencken, who observed: "Nobody ever went broke underestimating the intelligence of the American public." This is especially true about the issue of climate change.

Journalist Robert Samuelson[1] of the Washington Post Writers Group tersely summed up the problem in 2007 as more damning evidence became apparent about the decline of our natural world:

> Don't be fooled. The dirty secret about global warming is this: We have no solution. About 80 percent of the world's energy comes from fossil fuels (coal, oil, natural gas), the main sources of man-made greenhouse gases. Energy use sustains economic growth, which — in all modern societies — buttresses political and social stability. Until we can replace fossil fuels, or find practical ways to capture their emissions, governments will not sanction the deep energy cuts that would truly affect global warming.
>
> Considering this reality, you should treat the pious exhortations to 'do something' with skepticism, disbelief or contempt. These pronouncements are (take your pick) naive, self-interested, misinformed, stupid or dishonest. Politicians mainly want to be seen as reducing global warming when they're not. Companies want to polish their images and exploit markets created by new environmental regulations.
>
> Anyone who honestly examines global energy trends must reach these harsh conclusions.

---

1 Robert Samuelson, "Global Warming and Hot Air," *Washington Post*, Feb. 7, 2007.

Samuelson concludes: "It's a debate we ought to have — but probably won't. Any realistic response would be costly, uncertain and no doubt unpopular. That's one truth too inconvenient for almost anyone to admit."

Two trains — the Ecological Disaster Express and the Economic Engine That Could — are barreling along the same track in opposite directions. The crash is inevitable and the collision will be devastating.

After examining these global energy trends, I find that the conclusions are inescapable. A huge majority of the books on water scarcity, global warming, global political tensions, population growth, and other issues conclude with some plan of action, some way out, a prescription for earthly salvation presented gamely but in a noticeably half-hearted fashion. The resignation of those who understand the situation is evident, along with the profound sense of sadness — not so much for themselves but for the generations that will follow and for our complicity in the upcoming crises. I will exclude any natural proclivity to present a tidy solution in deference to the facts.

Ecologist Rachel Carson sounded the alarm in her book *Silent Spring*[1] in 1962 — over 45 years ago. Her reward was a no-holds-barred attack by corporate interests and their allies. The environment has deteriorated seriously since that time, and there is no reason to believe it will improve seriously in the future. The same powerful forces that went gunslinging for Carson are still with us. We've already wasted almost fifty years — and will, in all likelihood, waste the next twenty. The window will, by that time, be shut tight.

I have liberally used many sources in the course of writing this book and have quoted them extensively. The reason is simple. Most books on climate change view issues in isolation: scientists write about their particular fields of science, politicians about politics, sociologists about society, psychologists about psyches, economists about the economy, theologians and philosophers about religion, anthropologists about culture, engineers about infrastructure, and so on and so forth. This is a book aims to bring together these elements to create a synthesis, the complex whole: the effects of politics on science and energy policy, human behavior, the state of American society, international relations, and human interaction with the natural world. No single person is an expert in all of these fields. I have attempted to bring these various disciplines together by consulting the people who know and who, based on this knowledge, offer a view of what is likely to happen.

---

1 Rachel Carson, *Silent Spring* (Boston: Houghton Mifflin Company, 1962). A 40th Anniversary Edition was published by Houghton Mifflin in 2002.

CHAPTER ONE — WATER

> Once there were brook trout in the streams in the mountains. You could see
> them standing in the amber current where the white edges of their fins wimpled
> softly in the flow. They smelled of moss in your hand. Polished and muscular
> and torsional. On their backs were vermiculate patterns that were maps of the
> world in its becoming. Maps and mazes. Of a thing which could not be put back.
> Not be made right again. In the deep glens where they lived all things were older
> than man and they hummed of mystery.
> — *The Road*, Cormac McCarthy

> "The next World War will be over water."
> — Ismail Serageldin, vice-president of the World Bank

Water, while intricately involved in the cycle of life on the planet and drastically affected by climate change, deserves a primary and distinct focus. Humans are about 80 percent water and, unlike fossil fuels or even food, we can't live without it for long. Dehydration is a particularly unpleasant way to die. Water will play — and, in fact, is already a key factor — in geopolitics and an increasing cause of friction within and between countries. The United States is no exception. One flush of our toilets uses more water than a billion people are allotted per day.[1]

Fred Pearce, a former news editor at *New Scientist* and currently its environment and development consultant, sums up the situation this way:

> First the global picture. Higher air temperatures will increase evaporation from the world's oceans and so intensify the water cycle. By later this century, there could well be 8 to 10 percent more water vapor in the

---

1 Fred Pearce, *When the Rivers Run Dry: Water — The Defining Crisis of the Twenty-First Century* (Boston: Beacon Press, 2006).

atmosphere on an average day than there is today. That is 800 million acre-feet or so extra, enough to fill twenty Niles. This will almost certainly increase global rainfall. Also, shifting trajectories of rain-giving climate systems, like Atlantic cyclones, will mean that the total rainfall will be redistributed. Many middle latitudes will become drier. Meanwhile, the higher temperatures will also mean faster evaporation of water on land, so soils will dry out more quickly. This means less of the rainfall will reach rivers.

The rule of thumb seems to be that dry areas will become drier while wet areas will become wetter. Globally, climate will become more extreme, and rivers will respond in kind. As a result, many of the rivers that provide water in the world's most densely populated areas and where water is already in the shortest supply will be in still deeper trouble soon....[1]

The Dow Chemical Company, a once-notorious polluter trying to establish a more acceptable "green" public relations profile, published a map in *National Geographic*.[2] Despite the source, even a cursory glimpse at the map graphically depicts the problems — which are bad and getting worse. As Jared Diamond famously pointed out in his book *Collapse: How Societies Choose to Fail or Succeed*, most of the countries with severe water problems are also the most politically volatile.[3] Prime examples are North Africa and the Middle East. And with drought comes mass migration. Europe is expected to become the favored destination and a large number of Africans have fled to Spain, Italy, and France — countries already having problems assimilating the migrants. This has led to the creation of quasi-refugee camps to handle the flow. Both African and Middle Eastern countries have much higher than normal birth rates. With declining birth rates in Europe, most countries will need workers. With vastly accelerating birth rates, Muslims will need the work. To be blunt, the secular governments of Europe and the tenets of Islam have already created and are expected to increase high levels of cultural and religious tension. Hostility has already become evident in the United Kingdom, France, the Netherlands, and Denmark. More conflict is expected as recent polling demonstrates: most Europeans are highly suspicious of Islam and its potential effects on European culture.[4]

Today, a person whose annual water resources fall below 2,000 cubic meters runs the risk of joining the 700 million people in 43 countries living under the critical threshold of 1,700 cubic meters per person. At this level, water-stressed countries have difficulty meeting water requirements for agriculture, industry, energy, and the environment. By 2025 more than 3 billion people could be liv-

1 Ibid., p. 123.

2 Amy Corr, "*Dow Chemical Maps Out Plight of Unsafe Drinking Water*," MediaPost Publications, June 4, 2007.

3 Jared Diamond, *Collapse: How Societies Choose to Fail or Succeed* (New York: Viking Penguin, 2005).

4 "*Islam-West Riff Widens, Poll Says*," BBC News/Europe, Jan. 21, 2008.

ing in water-stressed countries — and 14 countries will slip from being water stressed to being water scarce (less than 1,000 cubic meters per person).

Water, of course, is used for many other purposes besides drinking. The main problem is water's role in agriculture; especially irrigation methods that bring it to areas that would otherwise be deserts. These regions and the humans living there would not be sustainable without irrigation. The problem is worldwide and the United States is not immune. The politics of the American West has always been the politics of water. The Colorado River and the Rio Grande are exhausted as states battle over water rights. Los Angeles should never have grown beyond a few hundred thousand people. During an average year, fewer than 10 inches of rain falls in Los Angeles. That's about one-quarter of the annual precipitation of a typical non-desert city like Boston.

"A desert, like Los Angeles, is by definition, literally, a place that isn't (or at least shouldn't be) inhabited.... The absence of water is what characterizes the desert's evolution and restricts its ability to sustain life," observes journalist Jeffrey Rothfeder in his book *Every Drop For Sale*.

Los Angeles is having an increasingly difficult time keeping up with its water problems. One element is population and another is the water supply itself: the number of people in the metropolitan area continues to skyrocket, but the city's sources of water are not providing as much as they used to. A 1964 Supreme Court ruling required L.A. to share a greater amount of the Colorado River — from which the city gets most of its water — with Arizona and Nevada, whose own populations are also increasing rapidly.

Southern California's desperate search for new water resources has forced it to turn to the private sector for solutions, as many municipalities around the world have done. Everywhere the worsening water equation is overwhelming the ability of governments and local authorities to provide creative solutions, while numerous for-profit companies have responded enthusiastically, suddenly seeing in water a commodity as potentially lucrative as oil, according to Rothfeder.[1]

Homes are being built in Arizona and Nevada where it is doubtful that a long-term water supply can be provided. Yet Easterners move to places like New Mexico and feel that life would be incomplete without a lush lawn and the water required for its maintenance.

In other parts of the world, Pakistan, for instance, cotton should not be grown — but it is, in abundance, and that will probably continue. The country consumes more than 40 million acre-feet of water a year from the Indus River — almost a third of the river's total flow and enough to prevent any water from reaching the Arabian Sea — just to grow cotton. Fred Pearce calculates that you could fill roughly 25 bathtubs with water needed to grow the 9 ounces of cotton in order

---

1 Jeffrey Rothfeder, *Every Drop For Sale: Our Desperate Battle Over Water in a World About to Run Out* (New York: Penguin Putnam Inc., 2001).

to make a T-shirt. Water must be put on crops in order to provide the food to feed an increasing growing population and Pearce's statistics are sobering: 250 gallons to grow a pound of rice; 130 for a pound of wheat; 65 for a pound of potatoes; 3,000 gallons to grow the feed for enough cow to make a quarter-pound hamburger; 500 to 1,000 gallons for that cow to produce a quart of milk. One pound of coffee? Ten tons. And the list goes on.[1]

The United States also exports about a third of all the water it withdraws from the natural environment, much of it through what is known as "virtual water" used to produce grains and meat.

The Middle East has already run out of water — the first major region to do so in the history of the world, according to Tony Allan of the School of Oriental and African Studies in London.[2] He estimates that more water flows into the Middle East each year as a result of imports of virtual water than flows down the Nile.

The Palestinian desert enclave of the Gaza Strip is the most water-starved political unit on Earth. Israel's need for water to maintain an essentially Western lifestyle is a little acknowledged issue in the seemingly intractable Middle East peace process. Others at the bottom of the hydrological heap include Kuwait and the United Arab Emirates, as well as island states like the Bahamas, the Maldives, and Malta.

If northern China were a separate country, it would be one of the most water-stressed nations on the globe.

All of this is important because of two changes over the past half-century: the soaring world population and the manner in which we have gone about trying to feed the population.

But hasn't the introduction of high-yield crops in the developing world eased the food shortage? These high yields come with a high hydrological price. These new crop varieties also need huge amounts of water now provided by dams and irrigation systems. Improvements in crop production have worsened the water problems. Today some 70 percent of all the water taken from rivers and underground reserves is being spread onto the 670 million acres of irrigated land that grows a third of the world's food. The granaries fill but the rivers empty. The world grows twice as much food as it did a generation ago, but it takes three times more water to do it. In arid countries such as Egypt, Mexico, Pakistan, and Australia and across Central Asia, 90 percent or more of the water extracted from the environment is used for irrigation.[3]

A quarter of India's crops are being grown using nonrenewable underground water. In some regions, salt from the irrigation water is invading the fields and rendering large areas sterile, losing 25 million acres a year. What was once a des-

---

1 Pearce, ibid., p. 4.
2 Professor Allan is also a member of the London Water Research Group, King's College, London.
3 Pearce, ibid., p. 233.

ert has been transformed by irrigation into farmland but is now returning to desert. India is increasingly sucking water out of both its rivers and its aquifers.

Individual farmers in India are relentlessly pumping water out of underground aquifers with pumps manufactured primarily in Japan. More than 21 million farmers tap underground reservoirs to water their fields. Aquifers, like oil deposits, build up over long periods of time and are not limitless. Arsenic has also been found in many of these wells. Underground water is invisible and only the local farmers know how much deeper they have to drill every year.

Without the Indus River, most of Pakistan would be a desert.[1] The river waters 90 percent of its crops and produces nearly half of its electricity. Pakistan is a major exporter of cotton and textiles which, as previously mentioned, consumes vast amounts of water. Crops are also being waterlogged and poisoned by salt. The population has quadrupled since independence in 1947 and is projected to reach 250 million by 2025. As salt builds up, it takes more and more water to enable fields to produce. The gloomy prognosis is that as Pakistan's population continues to soar, the country will have less and less land and water to produce the crops that sustain it. Water scarcity breaks down social cohesion. The population of Karachi has mushroomed to over 10 million with the majority living in slums which many fear could become a breeding ground for violence, perhaps both domestic and international.

As rivers fail, underground sources provide a third of the world's water. By some calculations, as much as a tenth of the world's food is being grown using underground water that is not being replaced by the rains. Major cities such as Beijing, Mexico City, and Bangkok are increasingly reliant on pumping out underground reserves. And the number of these megacities worldwide is expected to increase dramatically in the near future.

The Israelis have built dams on the *wadi* (water sources which are dry except in the rainy season). These days, little water ever reaches occupied Gaza and its Palestinian population. Most of it is grabbed to irrigate Israeli fields.[2]

Elsewhere, a multitude of dams have been erected during the 20th century, disrupting fish migrations and causing a drastic decline in wild fisheries. The world's rivers are no longer teeming with the fish that have always been a traditional food source. Increases in industrial pollution, thousands of new chemicals, garbage, and fuel spills have also fouled rivers, lakes, and oceans.

The desiccation of the Yellow River and its basin in China is an economic and environmental disaster. As its waters have faltered, millions of acres of farmland in the river's lower reaches have been abandoned. The wheat harvest has declined by a third and a dust bowl is a distinct possibility. If China cannot provide food for its population, it causes problems for the rest of the world. As China's

---

1 Ibid., p. 29.
2 Ibid., p. 62.

imports more and more grain, the world's granaries empty and global grain prices could soon rise by as much as a third. As the Yellow River empties, farms and cities try to keep going by pumping more groundwater. In some areas, inhabitants are pumping up water twice as fast as it is being replenished. In places around Beijing, 90 percent of the replenishable water is already gone.

The U.S. government's Scripps Institution of Oceanography estimates that reservoir levels in the Colorado River will fall by a third as declining rainfall and rising evaporation combine to reduce moisture by up to 40 percent across the Southern and Western states.[1]

Some rivers will run wilder, including those in northern Canada and Siberia which will dump more freshwater into the oceans. Many scientists believe that a large influx of freshwater will disrupt ocean currents which, in turn, will have a dramatic effect on the world's weather as the thermohaline conveyor belt ceases to operate. Alterations in the Gulf Stream would make temperatures in England and France, for instance, considerably colder.

In 2005, U.S. government climatologist Kevin Trenberth showed that after a century of little change, there has been a surge in instances of severe drought around the world since 1970. The proportion of the Earth's land surface suffering very dry conditions rose from the usual 15 percent to 30 percent at the start of the 21st century. Meanwhile, more freshwater keeps pouring into the Arctic Ocean, which intensifies global warming. This has been a trend since the 1960s.[2]

The massive glaciers of the Himalayas, Tibet, the Alps, and the Andes are expected to melt and rush down the rivers instead of acting as a gradual, predictable source of water for cities and towns located downstream. And then these waters will completely disappear. Floods will be followed by low or nonexistent river levels, and drought.

American researchers also believe that the spring meltwaters of the Sierra Nevada snow pack, which sustain summer irrigation of crops and lawns across the desert lowlands of California, could diminish 70 to 80 percent over the next 50 years. As the rushing floodwaters become a trickle, some of the most productive agricultural lands in the world could dry up and reduce global food supplies.

The role of dams in destroying the ecology has been extensive. Rather than "greening the desert," they have caused desiccated fields, salinization, and waterlogging. Silt has accumulated at the dams; it could have been used to create more fertile land downstream. The damming of rivers was all the rage in years past as countries and their leaders used the projects to project global status, modernity, and the ability to win the war of "man versus nature."

---

1 Tim Barnett and David Pierce, *"Climate Change Means Shortfalls in Colorado River Water Deliveries,"* Scripps Institution of Oceanography, University of California San Diego, April 20, 2009.

2 Pearce, ibid., p. 124.

A study published in 2005 found that the world's wild rivers are rapidly becoming extinct.[1] Of the 300 largest river systems, almost 200, including all of the 20 largest, now have dams on them. When it comes to hydroelectric production, all of the good sites have already been taken. You could fill every faucet in England for a year with the amount of water that evaporates annually from the surface of Egypt's Lake Nasser behind the Aswan High Dam.

In the American West, more than 6 feet of water evaporates annually from reservoirs like Elephant Butte on the Rio Grande, and Lakes Mead and Powell on the Colorado. A tenth of the flow of the Colorado River evaporates from Lake Powell alone. A typical reservoir in India loses 5 feet. Losses in the Australian outback can exceed 10 feet a year.

The World Commission on Dams warned that greenhouse gases were bubbling up from every reservoir that was measured. "There is no justification for claiming that hydroelectricity does not contribute significantly to global warming," the commission reported. Hydroelectricity, often viewed as a "green" renewable resource, is used extensively in Canada and parts of South America.[2]

As reservoirs age, most also produce substantial quantities of methane. Methane is about 20 times as potent as fossil-based CO_2 emissions. More dams would flood much more land for far less benefit and be significantly more expensive, according to Pearce.

Israeli hydrological rule over the West Bank since 1967 has been unyielding. Across most of the West Bank, Palestinians have been largely forbidden to sink new wells, and they rarely get permission to replace the old ones. Before the Israelis took control, the West Bank Palestinians had 774 wells. Thirty-five years later, only 321 were still operating. The rest either dried up or were off-limits by Israeli decree.

As the existing springs and wells deteriorate and their population grows, Palestinians find they have less water per capita than when the Israelis invaded. Today each Palestinian typically has less than a quarter as much water as his Israeli neighbor, a situation Pearce terms "hydrological apartheid." Some Palestinian families spend between 20 and 40 percent of their income to buy water. Another problem is Israel's security fence, which has cut off dozens of Palestinian villages from their traditional wells.

The Six-Day War has been called the first modern water war. And Israel's victory in seizing the Jordan River and its catchment remains an essential backdrop to the continuing conflicts. Israel today uses far more water than falls on its territory, and it has been able to do so because of its occupation of the West Bank, which gives it control of the western aquifer, and the Golan Heights, which gives it control of the Jordan River. "Peace talks cannot be understood without con-

1 Catherine Brahio, *"Dams Disrupt World's Major Rivers — Study,"* Independent Online, South Africa, April 19, 2005.
2 Pearce, ibid., p. 143.

sidering the water factor," Pearce comments. He added that much of the water is used to preserve Israel's Western lifestyle of lush lawns and daily showers.

Some people think the world's first nuclear exchange could one day be triggered between India and Pakistan because of events in Kashmir. But the spark may not be the suspects usually named in public — terrorism or border disputes. It could be water, according to Pearce. Kashmir is the gateway through which Pakistan receives most of its water along tributaries of the Indus River.[1]

Pakistan's 150 million people would be water-stressed without the Indus River. It runs the length of their country. Its waters irrigate most of their crops and generate half their electricity. But the Kashmir gateway is vulnerable. India and Pakistan have been in armed conflict three times since 1947, and the first conflict arose when India intervened in Kashmir to cut the flow of tributaries of the Indus. Some in Pakistan fear that India might intervene again. Countries always have and will go to war over resources. Add religious discord (between Muslim and Hindu) and nuclear weapons to the mix, and the results are many sleepless nights for world diplomats and security strategists.

Pakistan is a downstream state — at the mercy of others for its water. However, India accuses Pakistan of wasting most of the water and suffering no consequences for its overuse. Both Afghanistan and India contribute far more water to the Indus system than Pakistan but use much less.

Closer to home, the city of Phoenix is one of the biggest urban water users on the planet. Through the summer, residences in some suburbs consume around 1,000 gallons of water per day. But water tables are falling fast. Arizona now pumps up more than twice as much as the sparse rains can replenish. Like California, Arizona banked on future growth by tapping the Colorado River. The Central Arizona Project takes almost 1.6 million acre-feet of water a year out of the river and pours it into a concrete canal 300 miles long that zigzags across the desert to Phoenix and Tucson. In recent times, the canal has been taking more than a fifth of the entire flow of the Colorado — and much more in dry years. Its hydrological bounty was the engine for the real estate boom in Phoenix; that is now derailed, but it could have been the final straw for the Colorado River. And it could have taken down the American West with it.[2] Western states have a long history of battling over both the Colorado and the Rio Grande rivers.

While urban areas are taking an increasing amount, most of the water removed from the Colorado River still goes for irrigating some 4 million acres of fields in Arizona and California. America has always subsidized farming in the West, and today perhaps $1 billion a year is poured into keeping farmers irrigating crops that would not grow naturally. These subsidies encourage waste

1 Ibid., p. 75.
2 Ibid., p. 195.

and every year several million acre-feet of water evaporate from reservoirs, farm ponds, and flooded fields.

The U.S. Geological Survey recently stated that the Colorado River has not been so degraded in 500 years.[1]

Most observers of hydropolitics of the Colorado believe that the days when most of the waters of the river could be used to irrigate crops must come to an end in the near future. Expanding cities are demanding more water. And farmers are facing another threat — salt. Salt could literally kill off parts of the United States. As water expert Arthur Pillsbury noted before his recent death, "The Colorado basin will eventually become salt-encrusted and barren because of salt." As rivers dry up, salt can encroach and, as a result, cities can fall. Desert farming, artificially propped up by irrigation and vulnerable to salt, is not sustainable.

In the past 40 years, most of the Aral Sea in Central Asia has also turned into a giant desert. The United Nations has called the disappearance of the sea the greatest environmental disaster of the 20th century. Sixty percent of the rural population of Karakalpakstan drink from wells dug into the salty underground reserves in the desert. In many villages the water is so salty that milk added to tea instantly curdles. Mothers in some villages say their children will not take their breast milk because it contains too much salt.[2]

The threat of famine is returning to India after more than a decade in which the country has registered a food surplus. Southern India, like northern China, seems to be on the verge of a hydrological crisis. And with water tables tumbling and the country's population predicted to increase by 50 percent within the next 50 years, overtaking China to reach a staggering 1.5 billion people, there is a deep sense of impending crisis in India.

Across the developing world, raw sewage is increasingly used to irrigate crops. A study done by the International Water Management Institute concluded that perhaps a tenth of all the world's irrigated crops are watered by raw sewage from urban areas.[3] Without it, much of the world would go hungry. Along with the sewage come disease-causing pathogens laced with toxic waste from industry.

Desalination has often been viewed as a way to create more potable water from the oceans. But one problem is what to do with the salt extracted from the seawater. It emerges as a vast stream of concentrated brine. Maybe this can be solved one day. But what can't be solved is the huge energy demand of desalination, a topic which will be discussed later. Most desalination plants burn coal, oil, and other fossil fuels. So while desalination could conceivably become a viable source of drinking water in coastal regions around the world in the coming

---

1 *"Climate Fluctuations, Drought, and Flow in the Colorado River Basin,"* U.S. Geological Survey, Department of Interior, August 2004.

2 Pearce, ibid., p. 211.

3 *"Agriculture, Water and Cities,"* International Water Management Institute, Sri Lanka, Oct. 3, 2008.

decades, it would contribute to climate change. The desalination process is also expensive: it costs $2,000 or more per acre-foot, while in many regions water from lakes and rivers costs only half that. For this reason, it's being adopted so far only in countries that are near an ocean but have very little interior water and that have enough money (like Saudi Arabia) or at least the backing from international agencies to support this expensive operation.

Johan Rockstrom, a Swedish hydrologist, calculates that conquering existing hunger and coping with an estimated 3 billion additional people will increase global annual water consumption by 4.5 billion acre-feet, or 80 percent, by 2050. The ability to provide this water seems doubtful. Extra dams cannot come anywhere close to meeting that need, he says.[1] There's not enough money and, even if there were, most rivers in farming regions are already running dry and the good dam sites are mostly taken. At the same time, most of the world's underground water reserves are ruled out because they are already being overused and should, in fact, be scaled back. The future of the world's food may depend on a strategy from our ancient agricultural past — catching rain, according to Pearce.

By 2025, economists say, water scarcity will be cutting global food production by some 385 million tons per year. That is more than the current U.S. grain harvest and the equivalent of a loaf of bread every week for every person on the planet.[2]

An increasing number of irrigation canals are also drying up and underground reservoirs cannot take up the slack. We are already living on borrowed time by mining the aquifers — dipping into the slow, often largely non-renewable water cycle to accommodate the current increase in demand. Up to a billion people are today eating food grown using underground water that is not being replaced. Resources develop slowly — human consumption does not; it will only increase.

China is beginning to test its food limits. Historically, the world's most populous nation has almost always been able to feed itself. Today, increasing water shortages are pushing China to import large quantities of food for the first time. The country's population is already affecting world food security. Strategic grain reserves are emptying and world grain prices are rising. Meanwhile, the world's second most populous nation, India, is filling its granaries by plundering diminishing underground water resources, as Pearce and many others have observed. That is not sustainable.

While the world struggles to cope with myriad water problems, some entrepreneurs see opportunity. Water, after all, could be the next big thing — as many have said, water is poised to become the new oil. Like real estate, you can't make any more of it. In fact, water is even better than oil. People die, and die quickly, without it. This begs an ethical question: Is water a *right* or just another *commodity*

---

1 Pearce, ibid, p. 271.
2 Ibid, p. 306.

subject to the market forces of supply and demand? The debate has begun because the commodification of water has begun. Businesses are arguing that water is undervalued; people will waste more of it if use does not reflect its true cost. And if a few bucks are made in the bargain, all the better. More and more, water is not viewed as a shared natural resource that must be provided to everyone, but as a commodity that people need but are not inherently entitled to receive.

As the scramble for water goes on, there is a significant new development in water's world order: the emergence of a strong private market where water is being bought and sold as a commodity. With demand rising, supply dwindling, and traditional water management approaches producing more problems than solutions, the real value of water is increasing rapidly. Suddenly companies, like countries before them, see the potential for profits in the water trade, according to journalist Jeffrey Rothfeder.

As Rothfeder explains, "The aims of its participants can be described simply: to control, through physical and political management, the world's limited water supply. Whatever countries, companies, or people achieve this control will command the lion's share of power, capital, and influence in the world, especially as the amount of usable water per person dwindles during this century."

With demand for water increasing and the available amount dwindling, highly populated desert nations and communities are finding themselves increasingly defenseless. Just to import enough to survive, they are at the mercy of private companies and governments — which are, in turn, using water to strengthen their geopolitical position and increase profits.

The Middle East, the Earth's most densely populated desert region, is the center of some of the most complex and potentially dangerous jockeying for dominance and control of the area's water supply. Virtually every country in the region is running out of water quickly. With the region's population expected to increase by as much as 15 percent by 2025 — to about 350 million people — its water needs will be double what they were as recently as 1975. While the desert countries like Egypt, Syria, Iraq, Jordan, and Israel do have access to local groundwater, these sources are being depleted rapidly and won't be replenished in the near future. This water, found in underground aquifers, has been trapped since the last ice age and won't be renewed for thousands of years.

The name for Islamic laws — *shari'a* — stems from a word meaning "the sharing of water." In the seventh century, Islam emerged on some of the most barren landscape in the world. To spread the religion, Muslim leaders emphasized cooperation among the converts to defend against their enemies. The new religion, it was thought, would easily be destroyed if infighting broke out among the faithful over water rights. Water in the Middle East has always mattered.[1]

---

1 Rothfeder, ibid., p. 53.

"A time may well come when we have to calculate whether a war might be economically more rewarding than losing a drop of our water supplies," observed one leading Arab politician.

Subjecting water to free trade global policies and public-private partnership raises concerns. Companies are likely to neglect the environmental implications of their actions even more than governments; the price of water may become so high as to be prohibitive for poorer countries, which will suffer more than they do now; and decisions about water development contracts could be guided by political cronyism, enriching companies and government officials at the expense of the local population.

In what seems to defy logic, population growth in the inhabited deserts far outstrips that of the rest of the world. The world's deserts occupy nearly one-seventh of the Earth's land surface area, and most of them are unpopulated, which only makes sense — they have only 4 percent of the human population. Yet in the United States, the combined population of California, Nevada, and Arizona — the three states with the largest desert area — is expected to grow 62 percent by 2025. That is the largest projected increase of any region in the country. And every nation in the Middle East is experiencing a population boom that shows no sign of abating. This exponential population increase along with decreasing water resources is not sustainable.

Because of the growing need, water is rapidly becoming a for-profit commodity. Businesses have emerged to buy and sell it, and water projects are now designed to ultimately provide a return on investment. We are entering an age where the financial benefits of any given water deal will matter more than the needs of individuals.

For all the billions of dollars that the World Bank has provided for water development projects — about 14 percent of its overall funding budget since its inception — most of the benefits have accrued to multinational construction companies and the largest local industries. Very little — well under one percent of the bank's spending — has been allocated for providing water to average people.[1]

In developed nations — even in barren deserts like California and Nevada — water is virtually free to consumers. In the United States, a country where free markets are a virtual religion, such a low price for an item that is extremely limited in supply and high in demand is anathema to capitalist principles. If people who can afford water were charged a suitable rate based on free market conditions, and if a portion of this money were used to subsidize water delivery and water system construction for less wealthy countries, the water market itself would remedy a good deal of the crisis. Unfortunately, that is not the case.

One of Islam's main worries in the 7th century was that without a universal right to water, people would go to war over it, and thereby rip apart the new

---

1 Ibid., p. 90.

Muslim society. This pragmatic view is more relevant today than it was some 1,300 years ago. With our management of water and our neglect of the water crisis, it's an outcome that may manifest itself on a global scale.[1]

It's not surprising that globalization faced its first serious challenge, its first physical battle, in a struggle over water. Who owns the water in a community as well as how much clean water is delivered to residents and at what price speaks to the very basic issue of survival — unlike fruits, grains, or even petroleum.

Violent face-offs against the globalization of water may soon develop somewhere in the world. A complex series of water management decisions and new water management realities have made it almost a certainty. In the late 1990s, a handful of conglomerates began to quietly acquire control of the world's water systems. As the value of water began to soar along with the need for it, multibillion-dollar firms like Vivendi Universal, Suez Lyonnaise des Eaux, the now-defunct Enron, and Bechtel, among others, scoured the continents to, in some cases, purchase local water operations outright and in others to convince authorities to privatize their water concessions and let the companies run them.

It was a textbook example of what globalization has meant in practice — in this case, of the water industry — symbolized by large water corporations opening up new international revenue streams, frequently but not exclusively in developing countries and thus perpetuating past colonial resource grabs. In fact, water may be the biggest resource grab of all.[2]

Although it is only in its infancy, the private water sector in the United States generates more than $80 billion each year in revenue — four times Microsoft's sales. Worldwide, according to the World Bank, private water revenue globally approached $800 billion in 2000.

Governments, in both wealthy and poor countries, have often mishandled water supplies so badly for such a long time that their mistakes, poor planning, and ill-conceived policies are a prime reason that the amount of available water today is dwindling. Almost everywhere, the water infrastructure — the sewage pipes, the pumps, and the quality control stations — is falling apart, mostly out of neglect, and wasting vast quantities of water. And an increase in pollution is overwhelming the systems' ability to provide drinkable water.

There is less water available per person as the population grows. Private water companies argue that they're driven by free market rules (less supply, more demand) which assure a hefty profit. A literal revenue stream.

Privatization and commoditization have become so sanctioned as a fundamental solution to the water crisis that the International Monetary Fund, which provides loans to nations in trouble, frequently imposes water privatization (as well as many other forms of privatization) as a condition in its lending agreements.

---

1 Ibid., p. 51.
2 Ibid., p. 101.

Some of the poorest countries in the world, like Tanzania, Benin, and Rwanda, have been required to turn over local water systems to private corporations.[1]

Plans are being developed to transport water around the world and sell it to the highest bidder. Diverting more water from the few vital freshwater reservoirs that still exist would only worsen a bad situation, hastening the destruction of delicate ecosystems and local water supplies.

A focal point of contention in the United States is the Great Lakes region, which contains about 20 percent of the world's surface freshwater and, as such, is an obvious target as a source of water for overseas transfer and sale. This additional challenge comes as Great Lakes activists are already fighting an uphill battle against industrial pollution, agricultural runoff, and overuse. As many as 100 species and 31 ecological communities in the lakes are at risk of extinction, according to the Nature Conservancy. Environmentalists say that if large quantities of Great Lakes water are allowed to be shipped overseas, the level of the lakes would be lowered so severely that it would jeopardize their very survival. Toxins would begin to consume the shallow pools and strangle the essential plant life that trout, sturgeon, herring, and pike use for nesting and feeding. If that happens, the pollutants might seep into the water systems of dozens of communities that rely on the Great Lakes for their water supply. The particular problems of the Great Lakes will be examined later is this chapter.

Critics contend that water transfers would not only lower the levels in lakes and rivers, subjecting them to overwhelming amounts of toxic chemicals, but could also affect aquifers integral to local water systems. Aquifers depend in part on precipitation for replenishment and cleansing. Yet already upward of 50 percent of the world's aquifers are dangerously shallow or are becoming more and more contaminated with toxins. Environmentalists assert that in areas where freshwater is siphoned off for shipment to other parts of the globe, less water will be available for transpiration and thus less rain will occur to stock and purify aquifers. As a result, the amount of clean water in these aquifers could diminish to levels unable to sustain adequate local water supplies. In other words, to quench the thirst of people three thousand miles away, we could choke off the supply for the residents next door.

And water has become an infrastructure issue in the United States. In 2005, (and again in 2009) the American Society of Civil Engineers gave the drinking water infrastructure a grade of D-minus. According to the ASCE, "America faces a shortfall of $11 billion annually to replace aging facilities and comply with safe drinking water regulations. Federal funding for drinking water in 2005 remained level at $850 million, less than 10 percent of the total national requirement."[2]

---

1 Ibid., p. 115.

2 "*Report Card for America's Infrastructure*," American Society of Civil Engineers, Reston, Virginia, 2005, p. 2.

Business, not ecology, spurred action in Florida. The Everglades restoration project didn't come about because of environmental concerns over the destruction of a delicate and absolutely essential ecosystem. It occurred primarily because of financial fears — that Florida could not remain economically viable and attract new business and development if it was running out of water — and the sudden realization that turning on the tap in Florida and getting nothing out of it was becoming a possibility.

"If we look at today as a snapshot in time — that is, as if the problem of water scarcity were to hold at the current unacceptably high level long enough for us to do something about it — it would take a remarkable amount of technological skill, creativity, and financial resources to accomplish the goal of supplying enough water to enough people so that the crisis could be viewed merely as controllable," Rothfeder writes. But with every new misguided water project, we're adding the burden of billions of additional dollars and more and more difficult engineering feats to the solution, while dangerously telescoping the amount of time we have to implement one.[1] This could be a deadly combination for the planet, and it highlights the need to attempt what has been impossible until now: to consider water projects cautiously, instead of with abandon — meticulously analyzing them for environmental as well as economic consequences — before we make any more mistakes that we will be forced under pressure to try to undo.

Water can and has been used as a weapon. In 1999, Indonesian-backed militia opposing East Timor's independence killed thousands of guerrilla troops on the island and then purposefully threw the bodies in local water wells to pollute the aquifers. The Serbs in Kosovo, after a brutal rampage that murdered tens of thousands of Kosovar Albanians, dumped corpses into the province's wells, and within hours blood was pouring out of Kosovo's water faucets. And in a terrorist action in 1998, a guerrilla commander in Tajikistan planted a bomb at a dam on the Kairakkhum Channel, threatening to drown tens of thousands of central Asians if his demands weren't met.

In wealthy and poor countries alike, conflicts almost always erupt from the fear of not having enough clean water and very few regions in the world are free of this anxiety. If mounting demands from population growth, development, and industrial expansion outstrip supply, no amount of money can provide adequate water. Consequently, in the United States, for instance, states — and even people who live next door to each other within states — are increasingly bickering over shared water resources, usually to get a little more before it is all gone.

In most cases, these conflicts are fought in an intense political and economic atmosphere, where no one seriously discusses putting the brakes on population growth and corporate development — two factors that increase the need for more water. Only lip service is given to limiting sprawl, mandating conserva-

---

1 Rothfeder, ibid., p. 150.

tion, or managing water supplies more intelligently to eliminate the causes of the disputes. As a result, many fights over water in developed countries are never really settled — they're just buried for a time under a short-term fix and will inevitably emerge again as the amount of available water lessens and the need for it increases.

Americans can't claim they couldn't see the water crisis evolving. Writer Laurence Pringle pointed out the following problems over 25 years ago in his book *Water: The Next Great Resource Battle.* He made the following observations:

About a century ago, some people believed that much of the West was destined always to be lightly settled and little used. John Wesley Powell, famed for his exploration of the Grand Canyon, prepared a report on the arid West for Congress. He emphasized the natural limits of its water resources and of the land that could be irrigated. He concluded that "these lands will maintain but a scanty population."

Wrong.

"Today the arid West is booming in population growth and in agricultural and energy production," Pringle wrote. "Arid lands produce 66 percent of the nation's cotton, 39 percent of its barley, and 21 percent of its wheat. Southern California alone accounts for a sizable portion of the fruits and vegetables produced in the United States. Maps of the West are dotted with active or proposed energy projects, including uranium mining, oil and natural gas drilling, and coal mining. Most of the nation's fastest growing cities are in the arid West. All this is possible because of some technological devices and feats that Powell could not foresee — huge dams, powerful pumps, highly mechanized irrigation equipment, and systems that carry water hundreds of miles, from places of relative abundance to more arid regions. It is also made possible because only a small fraction of the water's cost is paid by those who use it; the blooming of the West has been financed largely by public funds from all of the nation's taxpayers."

Throughout Ogallala aquifer country, which stretches from Nebraska to Texas, some people must now face the disturbing idea that Powell was correct. Their wells have bottoms, and when the groundwater is mostly gone, the economy that depended on its extravagant use will also be gone.

As water needs increase in Colorado and other Upper Basin states, the Colorado River Compact seems more and more inequitable to residents of these states. In dry years especially, people in Colorado resent the idea that Californians are squandering water by washing cars, watering lawns, and filling swimming pools with Rocky Mountain water.[1]

"The Central Arizona Project was designed to fill about two-thirds of a huge groundwater overdraft in Arizona," according to Pringle. "Aquifers supply 62

---

1 Laurence Pringle, *Water: The Next Great Resource Battle*, (New York: MacMillan Publishing Co. Inc., 1982), p. 81.

percent of all water consumed in Arizona. As in other western states, the main use of groundwater — nearly 90 percent — is irrigation. In addition, the city of Tucson depends entirely on groundwater. It is the largest city in the United States to do so.

"Throughout Arizona, groundwater is being used twice as fast as it is being replenished. In Tucson, the rate is five times as fast. Water levels in some Tucson wells have dropped 110 feet in 10 years.

"Tucson residents were urged to save water, but this was a difficult adjustment for those who had once lived in the East or Midwest. To these people, a home without a lawn seemed incomplete. For them, trees, green lawns, and heavily watered golf courses helped to disguise the reality that Tucson is a desert, receiving only 11 inches of rain a year."

In California, while some farms are family-owned and operated, most irrigated land is controlled by such corporations as Tenneco, Union Oil, Getty Oil, Shell Oil, Standard Oil of California, and the Southern Pacific Railroad. A 1978 study in Kern County, north of Los Angeles, showed that 85 percent of all irrigation water was used by 15 large corporations. Consumer and environmental groups argued that giving more tax-subsidized water to such landowners was "socialism for the rich."[1]

In 1976, the General Accounting Office, which conducts studies for Congress, estimated that more than half of all irrigation water is wasted. The most wasteful methods — flooding whole fields or the furrows between rows — are still widely practiced. At best, half of the water reaches crop roots.

Rice and alfalfa are commonly grown on irrigated land in the Southwest. Both require large amounts of water. They don't need to be grown on arid lands. Under ordinary market conditions, water-consuming crops like these would be mostly grown in regions having abundant rainfall. They are being grown in the driest part of the United States because water is often priced at about 10 percent of its actual cost of delivery. Cheap water allows arid lands to compete with regions where water is delivered abundantly by rain.[2]

The Reagan Administration discouraged federal intrusion in so-called states' rights issues that affected water use. The federal government could have withheld funds for water projects until states controlled groundwater pumping and instituted water-saving irrigation methods. This never happened.

Dave Dempsey, who has been intimately involved in the ecological issues concerning the Great Lakes, has concentrated on the lakes' problems in his book *On the Brink: The Great Lakes in the 21st Century*.[3] For generations after European settle-

---

1 Ibid., p. 112.

2 Ibid., p. 127.

3 Dave Dempsey, *On the Brink: The Great Lakes in the 21st Century* (East Lansing, Mich.: Michigan State University Press, 2004), p. 162.

ment, Dempsey noted, the lakes were used as a dumping ground, "an open sewer which became a threat to public health."

Lake Erie fared worst of all. Through a combination of its nature and its extensive development, mostly on the United States side, the shallowest of the lakes was destined to become a global poster child for environmental disgrace on a developed continent. The virtual collapse of the Lake Erie ecosystem in the 1950s and 1960s represented the worst that democratic governments and capitalist economies could do to an ecosystem. As a child, I remember bobbing about the water in Lake Erie trying to avoid the massive fish kills that had become common. Only public outcries spurred the little action that has been taken to address the problems. The movement to reclaim the lake emerged in both the United States and Canada.

Treating individual symptoms in various lakes no longer seemed to work. First scientists, then advocacy organizations, and finally governments began to imagine the Great Lakes as one immense, complicated ecosystem. This was a breakthrough. Dempsey details the sources, nature and extent of the threats to the lakes' ecology, from high levels of pesticides, mercury, and other noxious chemicals to alien species including the zebra mussel, and the inability of politicians to go beyond symbolic politics and break ranks with industrial interests defending the status quo.

Beginning in the 1980s, politicians repeatedly and successfully crafted new laws and programs that sounded reassuring but failed to protect the lakes. Most of these initiatives proved to be riddled with exemptions, gaping loopholes, and either unenforceable or not intended to be fully implemented. Yet their enactment or proclamation would, for a time, satisfy a concerned public that something had been done.

The levels of the lakes, whose natural fluctuations have forced significant human adjustments and created bitter political controversies for over a century, will respond to climate change. A study commissioned by the U.S. government reported in 2000 that Great Lakes levels would drop "significantly" under most of the scenarios tested, with declines as great as two to five feet on Lakes Michigan and Huron by 2090.

If the idea of ecosystem integrity had been an operating principle for the governments of the Great Lakes region, climate change policies would have been far different. Instead, most of the region's governments ignored or even denied the existence of the problem. Particularly after the ascension to power in the United States of the George W. Bush Administration, which rejected the Kyoto climate change treaty and promoted research over action, inaction by the Great Lakes states was "an unwise gamble at best."[1]

---

1 Ibid., p. 233.

But we face an even greater threat: an urgent and worsening scarcity of fresh, as well as clean, water in the world. By 2025, up to three billion people in 52 countries are expected to have insufficient water to meet basic domestic and sanitation needs — a doubling from the year 2002. After listening to a presentation on looming water problems worldwide, International Joint Commission Canadian section chair Herb Gray said the talk "scared the hell out of me."[1]

These forecasts point to a grim future. The practices contributing to North American water scarcity are rooted in consumption patterns and freedoms held dear by U.S. and Canadian citizens. The remorseless tapping of groundwater resources to support agriculture and industry is drying up aquifers alarmingly fast, Dempsey noted. The Worldwatch Institute estimated in 1999 that 160 billion cubic meters of water were drained from groundwater each year, with the United States among the top five consumers.[2]

Enter Annex 2001, which placed an emphasis on local water use. Like other bold-sounding Great Lakes agreements before it, Annex 2001 was colliding with domestic political realities. It is one thing to rally against Sunbelt water pirates when defending the Great Lakes. But it is quite another to learn a water conservation ethic and to regulate one's own use. Living among or in sight of the world's largest lakes, residents of the Great Lakes Basin have seen no reason to limit their water use or consumption, and their governments were reluctant to educate them about the need for it. But now they were entering a new age of grave uncertainty and coming to grips with the vulnerability of those vast waters. As the International Joint Commission noted in its 2000 report: The waters of the Great Lakes are, for the most part, a non-renewable resource. They are composed of numerous aquifers that have filled with water over the centuries, waters that flow in the tributaries of the Great Lakes, and waters that fill the lakes themselves. Although the total volume in the lakes is vast, on average less than 1 percent of the waters of the Great Lakes is renewed annually by precipitation, surface water runoff, and inflow from groundwater sources, according to the report.

Lake Michigan and the other Great Lakes face an additional threat from another invasive species. Asian carp are making their way up the Illinois River, which connects the Mississippi River to Lake Michigan. The carp, imported by catfish farmers to remove algae from their ponds in the 1970s, are large and reproduce rapidly. They pose a significant risk to the Great Lakes ecosystem. Various agencies are working together to install and maintain a permanent electric barrier between the fish and Lake Michigan. The carp are voracious eaters, can weigh up to 100 pounds, and can grow to a length of more than four feet.[3] Whether or not this barrier will be effective is anyone's guess.

---

1 Ibid., p. 234.
2 Ibid., p. 235.
3 *"Invasive Species: Asian Carp and the Great Lakes,"* U.S. Environmental Protection Agency, Washington, D.C., Sept. 19, 2008.

The presidency of Barack Obama has brought new hope to those concerned about the Great Lakes but real action is needed to combat generations of abuse.

The U.S. House of Representatives recently authorized tripling federal money for Great Lakes toxic cleanups by a vote of 317-101.

The House authorized up to $150 million a year for the Great Lakes Legacy Act over the next five years, an increase from its current $54 million a year. However, the Senate still needs to act.

The vote followed President Obama's inclusion of $475 million for lake restoration in his proposed budget, which also includes funding for sewage work. Some foreign ships have also agreed to tougher regulations on ballast water which are believed to have introduced harmful species into the lakes.

According to author Jared Diamond, most of the world's freshwater in rivers and lakes is already being utilized for irrigation, domestic and industrial water, and other uses such as boat transportation corridors, fisheries, and recreation. "Throughout the world, freshwater underground aquifers are being depleted at rates faster than they are being naturally replenished, so that they will eventually dwindle. Of course, freshwater can be made by desalinization of seawater, but that costs money and energy, as does pumping the resulting desalinized water inland for use. Hence desalinization, while it is useful locally, is too expensive to solve most of the world's water shortages."

Today over a billion people lack access to reliable, safe drinking water.

There are by now literally hundreds of cases in which alien species have caused one-time or annually recurring damages of hundreds of millions of dollars or even billions of dollars. Lampreys have devastated the former commercial fisheries of the Great Lakes. An unfortunate but undeniable consequence of increased global trade has been the introduction of alien species with no natural predators.

While water is a renewable resource, its increasing scarcity, due to not only consumption but degradation through pollution and salinization, threatens the livelihood and security of states, with the shortage termed "water vulnerability." Rather than directly causing conflict, however, water scarcity tends to limit economic development, promotes resource capture, or leads to social segmentation, which in turn produces violence. Moreover, its transnational character, with rivers or underground aquifers crossing state borders, means that one country's use and actions affect neighboring states, according to K. Bruce Newbold, author of *Six Billion Plus*.[1]

While we could learn to live with less oil or precious minerals, we cannot live without adequate supplies of clean water and we are beginning to touch upon the limits of underground aquifers and even major rivers, according to Stephen M. Younger, author of *Endangered Species: How We Can Avoid Mass Destruction*

---

1 K. Bruce Newbold, *Six Billion Plus: World Population in the Twenty-First Century* (Lanham, Md.: Rowman & Littlefield Publishers, Inc., 2007), p. 198.

and Build a Lasting Peace. With the world's population expected to top 9 billion by 2050, there is serious concern over where we will get enough water for all these people. Relatively new technologies, such as desalinization of seawater, could work but only at a much higher cost than we pay today, affecting our ability to buy things and contribute to economic development around the world. Access to water could very well create a new class of haves and have-nots and a new set of domestic and international tensions. And the pressure will be even greater than it is for oil.[1] "Whereas oil is a necessity for industrial society, water is absolutely essential to all life," Younger writes.

According to David and Maria Pimentel in an essay *World Population, Food, Natural Resources, and Survival,* the minimum basic water requirements for human health, including drinking water, is considered to be about 50 liters per capita per day.[2] The U.S. average for all domestic usage, however, is eight times higher than that figure, or about 400 liters per capita per day. Americans use water freely in their agriculture, homes, and gardens. Most of the planet is not as fortunate.

Rapid human population growth and associated increased water demand already is stressing the world's water resources. Worldwide, between 1960 and 1997, the per capita availability of freshwater worldwide declined by about 60 percent. Another 50 percent decrease in per capita water supply is projected by the year 2025. Water demands already far exceed supplies in nearly 80 nations. For instance, in China more than 300 cities suffer from inadequate water supplies, and the problem is quickly intensifying as the population increases. In arid regions such as the Middle East and parts of North Africa, where yearly rainfall is low and irrigation is expensive, the future of agricultural production is grim.

Eroded soil sediments washed into the reservoirs behind dams during rainfall creates additional problems. Estimates are that about 1 percent of the volume of reservoirs is being filled with sediments each year, reducing the volume of water available for irrigation and other purposes. The total cost of sedimentation plus the loss of the water storage capacity of each dam worldwide is estimated to be about $7 million per year. It would be economically impractical to dig and remove the soil sediments deposited in the reservoirs.

According to Antonia Juhasz in her book *The Bush Agenda: Invading the World, One Economy At a Time,* lack of money and administrative skill in running public sectors has always been used as a reason for privatization. Corporate giant Bechtel

---

1 Stephen M. Younger, *Endangered Species: How We Can Avoid Mass Destruction and Build a Lasting Peace* (New York: HarperCollins Publishers, 2007), p. 115.
2 David and Marcia Pimentel, essay *World, Population, Food, Natural Resources, and Survival,* Global Survival: The Challenge and its Implications for Thinking and Acting, ed. Ervin Laszlo and Peter Seidel (New York: SelectBooks, Inc. , 2006), p. 35.

may well position itself as the only company with the ability to run the facilities it has already built, opening the door for its entrance as a water privatizer.[1]

"Why would Bechtel be interested in privatizing Iraq's water? After its oil, water is Iraq's second most valuable resource. Iraq is home to the most extensive river system in the Middle East, including the Tigris and Euphrates rivers and the Greater and Lesser Zab Rivers. It also has a sophisticated system of dams and river control projects. Stephen C. Pelletiere, a former CIA senior political analyst on Iraq during the Iran-Iraq war, has written that 'America could alter the destiny of the Middle East in a way that probably could not be challenged for decades — not solely by controlling Iraq's oil, but by controlling its water.' Even if America didn't occupy the country, once Mr. [Saddam] Hussein's Ba'ath Party is driven from power, many lucrative opportunities would open up for American companies; Bechtel has already become one such company."

A *Fortune* magazine article published in 2000 noted, "water promises to be to the 21st century what oil was to the 20th century: a precious commodity that determines the wealth of nations." Thus, the article concluded, "If you're looking for a safe harbor in stocks, a place that promises steady, consistent returns well into the next century, try the ultimate un-Internet play: water."[2] Investments in water systems therefore have the potential to be both highly lucrative and politically advantageous.

Author Pablo Rafael Gonzalez writes in his book *Running Out*: "The need for water is even more fundamental than the need for oil, and its shortage is perhaps more apparent, albeit often dismissed as merely a seasonal, regional or cyclical problem. There have always been regions where the availability of water has been poor and others where it has been plentiful; but in recent years, due to over-use of the underground aquifers and due to pollution of sources that once were pristine, even areas where traditionally water has been abundant now experience shortages.

> Water availability — renewable domestic water resources — in the United States and Canada amounts to 19.4 annual cubic meters per person; that means 18 times more than the availability for people in the Middle East and North Africa.

> As a United Nations report on world hunger once noted, "a hungry man is an angry man." A thirsty man is, too. The difficulty of obtaining water and food has not helped to calm things down in Africa and the Middle East, where there is little to go around and many mouths to feed."[3]

---

1 Antonia Juhasz, *The Bush Agenda: Invading the World One Economy at a Time* (New York: ReganBooks, 2006), p. 236.
2 "Is Water the New Oil?", *Fortune Magazine*, May, 2000.
3 Pablo Rafael Gonzalez, *Running Out: How Global Shortages Change the Economic Paradigm* (New York: Algora Publishing, 2006), p. 59.

The Jordan River has also been shrinking and this only intensifies conflict between Arabs and Jews.

Desert regions are not the only trouble spots. Glacier National Park in Montana is a fitting emblem for the great change sweeping the world's cold places. Dan Fagre has studied the glaciers in the park for 15 years. A scientist for the U.S. Geological Survey, he has the numbers at his fingertips: 27 glaciers left in the park out of 150 a century ago with 90 percent of the ice volume gone. He gives the remainder another 25 years. "It will be the first time in at least 7,000 years that this landscape has not had glaciers," according to Tim Appenzeller writing in the June 2007 edition of the *National Geographic*.

This lack of potable water will be a major factor — if not *the* major factor — in kick starting major crises in the 21st century.

CHAPTER TWO — OCEANS

> They trekked out along the crescent sweep of beach, keeping to the firmer sand below the tidewrack. They stood, their clothes flapping softly. Glass floats covered with a gray crust. The bones of seabirds. At the tide line a woven mat of weeds and the ribs of fishes in their millions stretching along the shore as far as eye could see like an isocline of death. One vast salt sepulchre. Senseless. Senseless.
>
> — Cormac McCarthy, *The Road*

We know about as much of the workings of the world's oceans as we know about deep space. Despite covering vast amounts of the Earth's surface, the oceans still contain as many mysteries as deep space or the ever-puzzling human brain — a triumvirate of yet-to-be explored nether regions. We have used the oceans mostly as a food source and garbage dump. Nature is beginning to take its revenge for this continuing assault. Oceans, however, could simplistically be described as the fan belt propelling the delicate engine of the planet. The belt has become increasingly frayed and could snap by the year 2030.

While climate change scientists have been attacked for their "alarmist" views by corporately-financed think tanks, the reality is quite the opposite — the deterioration of the world's oceans is accelerating at a pace that was woefully *underestimated* by the International Panel on Climate Change (IPCC).

"The Arctic is screaming," Mark Serreze, senior scientist at the government's snow and ice data center in Boulder, Colo., told the Associated Press in 2007.[1] Some scientists speculate that we are now beyond the ominous tipping point in our race to keep up with climate change. The computer models are consistently

---

1 *"Arctic Sea Ice Drops to Second Lowest Level,"* The Associated Press, Aug. 28, 2008.

underestimating the rate of melting ice in the Arctic. Summer sea ice could disappear in as little as five years.

"The Arctic is often cited as the canary in the coal mine for climate warning," said NASA climate scientist Jay Zwally. "Now as a sign of climate warming, the canary has died. It is time to start getting out of the coal mines."[1]

Climate change is notoriously hard to predict, let alone prove, because of the mind-boggling amount of variables that affect weather. Climate change deniers have skillfully exploited this complexity by demanding definitive proof. As most people are well aware, weather forecasts are virtually worthless beyond a three-day period for exactly this reason. But there can be no doubt that the air-conditioning system of our planetary home is rattling badly. And there is a strong smell of gas in the kitchen.

It would be prudent to repair the air conditioner before its total collapse. It would not be prudent to strike a match in the kitchen. The problems will become unfixable because our global home has no owner's manual. In what could only be termed a public relations coup, short-term profiteers have tarred scientists as a group of "Chicken Littles" spreading gloom and doom where none exists. Both within the United States and throughout the world, deniers have duped their constituents to act against their own best interests. When the problems become evident, it will already be too late. It's simple. Capitalism cannot survive without short-term growth; the planet cannot sustain this growth without collapsing. The two are antithetical. We cannot save the planet — and the human race — by ever-expanding growth directed by the "invisible hand" of free markets and globalization. The forces of capitalism are too strong to be defeated and the radical change required by societies will never be implemented because it will simply cause too much pain. It *cannot* be business as usual but it *will be* business as usual.

The greatest amount of ecological damage — by far — has come from the children of the 1960s, the baby boomers, through an orgiastic spree of consumption that contradicted any feigned concern for future generations. Boomers will die relatively unscathed by climate change; the next generation will not. While baby-boom parents are loath to allow their children to ride a bicycle without a helmet, they have been extremely cavalier about their long-term prospects for survival.

Any rock 'n' rolling baby boomer should be familiar with the concept of "feedback," the ear-splitting unintended consequences of amplification. It's a buzz-killer. Unintended consequences will also affect the oceans and, hence, the global ecological system. Melting sea ice is becoming one part of a troubling spiral.

White sea ice reflects about 80 percent of the sun's heat off of the Earth. When there is no sea ice, about 90 percent of the heat goes into the ocean, which then warms everything else. Warmer oceans then lead to more melting.

---

1 Ibid.

"The feedback is the key to why models predict that the Arctic warming is going to be faster," Zwally said. "It's getting worse than models predicted."

It's Henry Kissinger's famed domino theory about the spread of communism. Kissinger was wrong about his political theory, but it becomes much more relevant when applied to the oceans. One falling domino fells another which, in turn, fells another, *ad infinitum*, until the system breaks down irrevocably. The fan belt breaks. There are no replacement parts.

Science is hard; ignorance is bliss. Scientists are not very effective when it comes to explaining complex problems to a wary and easily-bored general public. The scientific information must pass through a political filter where its effectiveness fails miserably. These crises were anticipated at least 50 years ago followed by half a century of inaction.

How bad is it? Ask Dr. Peter D. Ward, a professor of biology and earth and space sciences at the University of Washington. Author of *Under A Green Sky: Global Warming, the Mass Extinctions of the Past and What They Can Tell Us About Our Future.* He also serves as an astrobiologist with NASA.[1]

> In the early 1990s a large mass of warm, low-oxygen water would rise from the depths and kill all the corals of the Short Drop-Off, even those in the shallowest water. The lethal deepwater was very warm, that warmth having been generated by Earth's global warming. Today, like so many reefs around the world, the once thriving reef community at Palau's Short Drop-Off is a cemetery ultimately caused by anthropogenic (man-made) carbon dioxide, a victim of what came to be known as coral bleaching, thanks to the washed-out colors it and other reefs would develop as they succumbed to water too warm. It would be one of the first sirens announcing an oncoming greenhouse extinction, if my colleagues and I have correctly interpreted the clues from the past. The time for studying the nautiluses came and went, another decade passed, and with increasing heat the reefs began to die. *Something Wicked This Way Comes*, to steal a phrase from Ray Bradbury.[2]

Dr. Ward discusses holes in the atmosphere and their effects on the food chain. Especially in the Antarctic, the biomass of photoplankton rapidly decreases under such holes. (In fact, in late 2006, the hole over Antarctica was the largest ever observed.) If the base of the food chain is destroyed, it is not long until the organisms higher up the food chain also suffer.[3]

Oceanic currents play a huge role in climate and global temperature. Today, the conveyer-current system in the north Atlantic Ocean seems to be controlled by the amount of ice cover on Earth, and is influenced by tropical warming or cooling, according to Ward. Scientists now have worked out the mechanisms underlying the conveyer-current systems. Ward says it is time to stop looking at

---

1 Peter D. Ward, *Under A Green Sky: Global Warming, the Mass Extinctions of the Past and What They Can Tell Us About Our Future* (New York: Collins, 2007), p. 114.
2 Ibid., p. 114.
3 Ibid., p. 118.

the "kill mechanisms" — low oxygen, heat, and perhaps excess hydrogen sulfide gas in water and air — and start looking at the driver of these changes — the atmosphere itself.

Ward and others believe that it is "pretty clear" that times of high carbon dioxide — and especially times when carbon dioxide levels rapidly increased — coincided with mass extinctions. This is the driver of extinction.

Researcher Minze Stuiver and many others working on the ice-core record showed that 200,000 and 10,000 years ago, the average global temperature had changed as much as 18 degrees Fahrenheit in a few decades. The average global temperature was 59 degrees Fahrenheit in 2003.[1] "Imagine that it suddenly shot to 75 degrees Fahrenheit or dropped to 40 degrees Fahrenheit, in a century or less," Ward says. "We have no experience of such a world and what it would be like; such sudden perturbations in temperature would enormously alter the atmospheric circulation patterns, the great gyres that redistribute the Earth's heat. At a minimum, such sudden changes would create catastrophic storms of unbelievable magnitude and fury. Yet such changes were common until 10,000 years ago." According to new studies from Greenland, it appears that the sudden shifts in the weather then disappeared. The salient point is that this respite is the exception and not the norm.

Soon after the start of this calm, humans began to build villages and cities, learned to smelt metal and conquer nature. And most importantly, humans learned how to tame crops and domesticate animals. Human population numbers began to soar as larger mammals underwent wholesale extinction. But there were still a few climate bumps in the road for both humans and animals, for the climate has had a history of rapid change. How rapid were these changes? The answer, according to Ward, is unsettling. Ice-core samples indicate that a global temperature change of 10 degrees Fahrenheit could take place in as little as 10 years.

Ward believes we may be seeing the start of a changeover that has now been recognized as having happened repeatedly up to 8,000 years ago and then stopped. The conveyer system in its present state has been stable for the amount of the time that humans have practiced agriculture, and this stability has allowed both predictability of crop yields in Europe and Asia, as well as the biologically more important stability of ecosystems. Ecologists have long known that organism diversity rises with stability. It is rapid change that leads to loss of biomass and biodiversity, ending with mass extinction. Ward cites and explains numerous studies detailing the science behind these phenomena and outlines how humans will be affected by loss of habitat and crop land, disease, violent weather and other downstream effects.

---

1 Ibid., p. 146.

But Ward cites the greatest threats caused by global warming as famine and war. People always did — and always will — fight over resources. Nations are unlikely to allow their populations to starve or their national treasures be depleted in order to buy enough food and other essentials. It will become more and more tempting to simply take them.

It would seem that plants might flourish in the higher carbon dioxide levels, and with longer growing seasons, perhaps an additional crop could be counted on in many areas. This is a benefit often cited by climate change skeptics, the "Won't It Be Great to Have a Summer Home in Newfoundland" argument. This might be true for some tropical fruits and starches. But the staple of human sustenance, grains and cereals, the very first crops from 10,000 years ago, would suffer. The grain belts rely on cool but not frigid winters and summers with abundant moisture. Current projections are that the great breadbaskets of Earth, especially the greatest of them all, the American Midwest, would have climate changes that would reduce summer moisture. As droughts become more frequent, yields of wheat, corn, barley, and oat crops would decline.[1]

Even if the human species is not endangered, many animals and plants will be. Ten percent of all species on Earth went extinct after previous greenhouse extinctions.[2]

About one-third of the world's coral reefs — the oceanic equivalent of tropical rainforests because they are home to a disproportionate fraction of the ocean's species — have already been severely damaged. If the trends continue, about half of the remaining reefs would be lost by the year 2030, according to Jared Diamond in his book *Collapse*.

However, some rare good news about coral reefs was reported in 2009.

Some Pacific Island countries are successfully protecting their coral reefs, haddock and scallops are recovering in New England waters, and a few types of whales are even making a comeback.[3]

"The news today is that there is good news" for the oceans, Nancy Knowlton of the Smithsonian's National Museum of Natural History said at a 2009 meeting of the American Association for the Advancement of Science held in Chicago. But she stressed that it doesn't mean that people no longer need to be concerned about the future of the oceans or sea life, but said it was time to move beyond the obituaries and recognize progress.

Australia has protected much of the Great Barrier Reef, and areas of the Northern Line Islands in the Pacific Ocean retain healthy reefs with large population of living coral, Jeremy C.B. Jackson of the Scripps Institution of Oceanography said.

---

1 Ibid., p. 190.
2 Ibid., p. 204.
3 *"Ocean Life Beginning to Register a Comeback,"* The Associated Press, Jan. 26, 2009.

Coral communities protected from overfishing are better able to resist global warming, Jackson added. He said they may bleach like other corals, but the protected ones bounce back.

Jackson also said he was encouraged by President Obama's intended policies.

"It's going to take a while, because this is a serious business, but the commitment is there in this administration," Jackson added.

Developing countries don't always have the luxury of protecting their fishing areas, commented Joshua E. Cinner of James Cook University in Australia. But he added that some succeed, using such measures as closing rotating areas to fishing.

The Northwest Passage was expected to be ice-free year round by 2050 but this now seems optimistic as reality again outraces computer models.

Sixty percent of the world's coral reefs are expected to disappear in the next few decades. The vast majority are already threatened or dead.

Canada's Atlantic cod fishery provides another example of the depletion of a renewable resource to the point where its future is questionable. Once dominated by small, individual fishermen and so plentiful that fishermen could catch the cod in buckets, the fishery was self-sustaining. Improved technology, including the introduction of offshore trawlers and factory ships, allowed increased catches of cod beyond a level that the fishery could support.[1] "While the cod fishery fell victim to overfishing, it also lacked awareness of the threshold of sustainability of the industry, as well as ignorance of the reproductive cycle of the codfish and an inability to forecast stock sizes correctly," according to K. Bruce Newbold in the book *Six Billion Plus.*[2]

The researchers of a peer-reviewed study by the National Academy of Sciences released in October of 2007 stated that the world's oceans are having more difficulty absorbing increased human emissions of carbon dioxide. The report said that human-induced warming has caused changes in wind patterns over the Southern Ocean that brought carbon-rich water toward the surface, reducing the ocean's ability to absorb excess CO_2 from the atmosphere. Alan Robock, associate director of the Center for Environmental Prediction at Rutgers University, added that "what is really shocking is the reduction of the oceanic CO_2 sink" — meaning the ability of the ocean to absorb carbon dioxide, removing it from the atmosphere. Robock said he thinks rising ocean temperatures reduce the ability of the oceans to take in carbon dioxide.

"It turns out that global warming critics were right when they said that global climate models did not do a good job at predicting climate change," Robock said. "But what has been wrong recently is that the climate is changing even faster than

---

1 Newbold, ibid., p. 181.
2 Ibid., p. 181.

the models said. In fact, Arctic sea ice is melting much faster than any models predicted, and sea level is rising much faster than the IPCC previously predicted."

The *Scientific American* also noted this problem cited by an Australia-based research consortium that analyzes the pollution behind global warming. The world's oceans and forests are already so full of CO_2 that they are losing their ability to absorb this climate change culprit. New maritime measurements over the past decade also show that the North Atlantic's ability to absorb CO_2 has been cut in half, according to researchers from the University of East Anglia who were not affiliated with the Global Carbon Project, the Australian group.

Carbon dioxide is dissolving in ocean waters and turning them more acidic, according to the November 2007 issue of *National Geographic*. This is bad news for shell-building animals because it can create a corrosive, even deadly, environment. Oceans are a natural sink for CO_2, already soaking up more than a quarter of what's released into the atmosphere. The oceans are now taking in 25 million tons a day of excess CO_2. Already scientists have measured a rise in acidity of some 30 percent in surface waters, and they predict a 100 to 150 percent increase by the end of the 21st century. [1]

No ill effects have been documented so far in the open ocean but the threat is clear. Absorbed by seawater, CO_2 reacts to form carbonic acid, which turns the normally alkaline water more acidic. In the process, fewer carbonate ions are left floating around — and many marine organisms, including mollusks and corals, rely on carbonate from seawater to build their shells and other hard parts. Eventually, vital species will no longer be able to build or maintain their shells and skeletons.

Users of the mineral aragonite — a very soluble type of calcium carbonate — are especially vulnerable. They include tiny pteropod snails, which help feed commercially vital fish like salmon. Computer models predict that polar waters will turn hostile for pteropods within 50 years (cold water holds the most CO_2, so it is already less shell-friendly). Impacts will be felt up the food chain. And as the acidification reaches the tropics, "it's a doomsday scenario for coral reefs," says Carnegie Institution oceanographer Ken Caldeira. If current trends continue, he predicts, reefs will one day survive only in walled-off, acid-controlled refuges.

Massive outbursts of CO_2 and other greenhouse gases have acidified the oceans in the geologic past, but equilibrium returned as the oceans stored away excess CO_2 in minerals on the seafloor. This time nature may be slow to heal. "Our emissions are huge compared with natural fluxes," Caldeira said. "If you could stop emissions and wait 10,000 years, natural processes would probably take care of most of it." We are producing more CO_2 than the oceans can handle.

---

1 *"The Acid Threat: As CO_2 Rises, Shelled Animals May Perish,"* National Geographic, November 2007.

In a *National Geographic* story by Tim Appenzeller in June 2007, he quotes Eric Rignot, a scientist at NASA's Jet Propulsion Laboratory who has measured a doubling in ice loss from Greenland over the past decade. Rignot said, "We see things today that five years ago would have seemed completely impossible, extravagant, exaggerated."[1]

Millions of people in countries like Bolivia, Peru, and India who now depend on meltwater from mountain glaciers for irrigation, drinking, and hydropower could also be left high and dry.

The temperature threshold for drastic sea-level rise is near, but many scientists think we still have time to stop short of it, by sharply cutting back consumption of climate-warming coal, oil, and gas. Few doubt, however, that another 50 years of business as usual will take us beyond a point of no return. Because the true effects of climate change show up many years after the damage has already been done, other scientists believe that it is imperative to begin radical global action no later than the year 2015. With the current global meltdown and the enormous costs of combating climate change, this goal is beginning to seem more and more like a pipe dream.

There will be virtually nothing left to fish from the seas by the middle of this century if current trends continue, according to one major scientific study. Stocks have collapsed in nearly one-third of sea fisheries and the rate of decline is accelerating.

Writing in the journal *Science*, an international team of researchers said fishery decline is closely tied to a broader loss of marine biodiversity. But a greater use of protected areas could safeguard existing stocks.

"The way we use the oceans is that we hope and assume there will always be another species to exploit after we've completely gone through the last one," said research leader Boris Worm, from Dalhousie University in Canada. "What we're highlighting is there are a finite number of stocks; we have gone through one-third, and we are going to get through the rest."

Steve Palumbi, from Stanford University in California, one of the other scientists on the project, added: "Unless we fundamentally change the way we manage all the ocean species together, as working ecosystems, then this century is the last century of wild seafood."

This vast piece of research, incorporating scientists from many institutions in Europe and the Americas, is drawing on four distinctly different kinds of data. Catch records from the open sea give a picture of declining fish stocks, according to Richard Black, environmental correspondent for BBC news.[2] In 2003, 29 percent of open sea fisheries were in a state of collapse, defined as a decline to less than 10 percent of their original yield. Bigger vessels, better nets, and new

---

1 Tim Appenzeller, *"The Big Thaw,"* National Geographic, June, 2007.
2 Richard Black, *"Only 50 Years Left for Sea Fish,"* BBC News, Nov. 2, 2006.

technology for spotting fish are not bringing the world's fleets bigger returns — in fact, the global catch fell by 13 percent between 1994 and 2003. Historical records from coastal zones in North America, Europe, and Australia also show declining yields, in step with declining species diversity; these are yields not just of fish, but of other kinds of seafood too. Zones of biodiversity loss also tended to see more beach closures, more blooms of potentially harmful algae, and more coastal flooding.

"The image I use to explain why biodiversity is so important is that marine life is a bit like a house of cards," Dr. Worm said. "All parts of it are integral to the structure; if you remove parts, particularly at the bottom, it's detrimental to everything on top and threatens the whole structure. And we're learning that in the oceans; species are very strongly linked to each other — probably more so than on land."

Protecting stocks demands the political will to act on scientific advice — something which Dr. Worm finds lacking in Europe, where politicians have ignored recommendations to halt the North Sea cod fishery year after year. Without a ban, scientists fear the North Sea stocks could follow the Grand Banks cod of eastern Canada into apparently terminal decline.

"It's amazing, it's very irrational," Dr. Worm told the BBC. "You have scientific consensus and nothing moves. It's a sad example; and what happened in Canada should be such a warning, because now it's not coming back."

The projected global collapse of ocean fish stocks by the year 2048 grabbed headlines when *Science* magazine published the alarming findings of an international panel of scientists, according to the ocean advocacy group SeaKeepers. While that report focused on the larger, fish species that we typically eat, equally important is the bottom of the complex ocean food chain. As with any structure, its stability and strength are based upon its foundation. Plankton and krill are the small animals and plant life that make up the base of the ocean food chain.[1]

Plankton are extremely small, free-swimming organisms that can be single- or multi-celled. There are two major plankton categories: phytoplankton and zooplankton. As the names suggest, phytoplankton are tiny algae or plants, and zooplankton are the animal counterparts. Krill consist of a group of about 85 separate species of shrimp-like animals up to two inches in length; the Antarctic Krill is one of the most significant.

By far, plankton and krill collectively are the largest biomass on the planet, with an estimated 150 million tons of krill alone spawned in the oceans annually. In comparison, the total world consumption of fish and shellfish is approximately 100 million tons a year.

---

1 *"Plankton and Krill Levels Rapidly Decreasing,"* The International SeaKeepers Society, spring 2007.

Plankton are a key component of the marine ecosystem and are sensitive to environmental change. Many natural resources directly depend on plankton as a food source. Not only is plankton the diet of small krill, but of the largest animal on Earth — the blue whale. Plankton provide the link between the atmosphere and the ocean as they pass energy of sunlight and nutrients and is the food web to fish, bird, and mammal populations.

Scientists have now correlated the increase in ocean temperatures with a decrease in phytoplankton production. The ocean is warming. In fact, 80 percent of the heat that is being trapped by greenhouse gases is absorbed by the ocean.

In some areas of the ocean, there has been a 30 percent decrease in phytoplankton production between 1999 and 2004 alone. These small algae convert or sequester a tremendous quantity of carbon, consumed in the form of the greenhouse gas carbon dioxide. The 30 percent reduction in phytoplankton would equate to about 190 million tons of unabsorbed carbon per year — a significant quantity in the global carbon equation.

Krill feed on phytoplankton near the ocean surface at night, but sink deeper in the water column during the day to hide from predators. New research shows the importance of these small animals in balancing greenhouse gases. It was recently discovered than Antarctic krill absorb substantial amounts of carbon; when they sink they then carry it down and bury it in the deep ocean.

In a recent issue of the journal *Current Biology*, scientists from the British Antarctic Survey (BAS) and the Scarborough Centre of Coastal Studies at the University of Hull discovered that Antarctic krill "parachute" from the ocean surface to deeper layers several times during the night.[1]

Lead author Dr. Geraint Tarling from the BAS said, "We've known for a long time that krill are the main food source for whales, penguins, and seals, but we had no idea that their tactics to avoid being eaten could have such added benefits to the environment. By parachuting down they transport carbon, which sinks ultimately to the ocean floor — an amount equivalent to the annual emissions of 35 million cars — and this makes these tiny animals much more important than we thought."

But understanding the correlations between global warming and marine ecosystems is complex; there are still anomalies and much more relevant, accurate information is needed. Collecting large amounts of quantifiable data, such as ocean temperature, pH, dissolved oxygen, chlorophyll, and nutrients, is crucial in order to refine our understanding of what is happening to the krill and plankton.

According to NASA scientist James Hansen: "Arctic sea ice is an example of a tipping point in the climate system.[2] As the warming global ocean transports more heat into the Arctic, sea ice cover recedes and the darker open ocean surface

1 *"Satiation Gives Krill That Sinking Feeling,"* Current Biology, vol. 16, issue 3, Feb. 6, 2006.
2 Alister Doyle, *"Arctic Thaw May Be at 'Tipping Point,'"* Reuters, Oct. 1, 2007.

absorbs more sunlight. The ocean stores the added heat, winter sea ice is thinner, and thus increased melting can occur in following summers, even though year-to-year variations in sea ice area will occur with fluctuations of weather patterns and ocean heat transport.

"Arctic sea ice loss can pass a tipping point and proceed rapidly. Indeed, the Arctic sea ice tipping point has been reached. However, the feedbacks driving further change are not 'runaway' feedbacks that proceed to loss of all sea ice without continued forcing. Furthermore, sea ice loss is reversible. If human-made forcing of the climate system is reduced, such that the planetary energy imbalance becomes negative, positive feedbacks will work in the opposite sense and sea ice can increase rapidly, just as sea ice decreased rapidly when the planetary energy imbalance was positive.

> Planetary energy imbalance can be discussed quantitatively later, including all of the factors that contribute to it. However, it is worth noting here that the single most important action needed to decrease the present large planetary imbalance driving climate change is curtailment of CO_2 emissions from coal burning. Unless emissions from coal burning are reduced, actions to reduce other climate forcings cannot stabilize climate.

> The most threatening tipping point in the climate system is the potential instability of large ice sheets, especially West Antarctica and Greenland. If disintegration of these ice sheets passes their tipping points, dynamical collapse of the West Antarctic ice sheet and part of the Greenland ice sheet could proceed out of our control. The ice sheet tipping point is especially dangerous because West Antarctica alone contains enough water to cause about 20 feet of sea level rise.

> Hundreds of millions of people live less than 20 feet above sea level. Thus the number of people affected would be 1,000 times greater than in the New Orleans Katrina disaster.... [R]epercussions would be worldwide.

> Ice sheet tipping points and disintegration necessarily unfold more slowly than tipping points for sea ice, on time scales of decades to centuries, because of the greater inertia of thick ice sheets. But that inertia is not our friend, as it also makes ice sheet disintegration more difficult to halt once it gets rolling. Moreover, unlike sea ice cover, ice sheet disintegration is practically irreversible. Nature requires thousands of years to rebuild an ice sheet. Even a single millennium, about 30 generations for humans, is beyond the time scale of interest or comprehension to most people."

Hansen said that on millennial time scales, sea level, CO_2, and global temperature change together.

> However, close examination shows that sea level has been stable for about the past 7,000 years. In that period the planet has been warm enough to prevent an ice sheet from forming on North America, but cool enough for the Greenland and Antarctic ice sheets to be stable. The fact

that the Earth cooled slightly over the past 8,000 years probably helped to stop further sea level rise.

Sea level stability played a role in the emergence of complex societies. Day and others pointed out in 2007 that when sea level was rising at the rate of 1 meter per century or faster, biological productivity of coastal waters was limited. Thus it is not surprising that when the world's human population abandoned mobile hunting and gathering in the Neolithic [12,000-7,000 years ago] they gathered in small villages in foothills and mountains. The study also noted that within 1,000 years of sea level stabilization, urban ... societies developed at many places around the world. With the exception of Jericho, on the Jordan River, all of these first urban sites were coastal, where high protein food sources aided development of complex civilizations with class distinctions.

Modern societies have constructed enormous infrastructure on today's coastline. More than a billion people live within 25 meter elevation of sea level. This includes practically the entire nation of Bangladesh, almost 300 million Chinese, and large populations in India and Egypt, as well as many historical cities in the developed world, including major European cities, many cities in the Far East, all major East Coast cities in the United States, among hundreds of other cities in the world."

How much will sea level rise if global temperature increases several degrees? Hansen said:

Our best guide for the eventual long-term sea level change is the Earth's history...The last time the Earth was 2-3 degrees Celsius warmer than today, about 3 million years ago, sea level was about 25 meters higher. The last time the planet was 5 degrees Celsius warmer, just prior to the glaciation of Antarctica about 35 million years ago, there were no large ice sheets on the planet. Given today's ocean basins, if the ice sheets melt entirely, sea level will rise about 70 meters or about 230 feet.

The main uncertainty about future sea level is the rate at which ice sheets melt. This is a "nonlinear" problem in which positive feedbacks allow the possibility of sudden ice sheet collapse and rapid sea level rise. Initial ice sheet response to global warming is necessarily slow, and it is inherently difficult to predict when rapid change would begin. I have argued that a "business-as-usual" growth of greenhouse gases would yield a sea level rise this century of more than a meter, probably several meters, because practically the entire West Antarctic and Greenland ice sheets would be bathed in meltwater during an extended summer melt season.

The Intergovernmental Panel on Climate Change calculated in 2007 a sea level rise of only 21-51 centimeters by 2095 for "business-as-usual" scenarios, but their calculation included only thermal expansion of the ocean and melting of alpine glaciers, thus omitting the most critical component of sea level change, that from ice sheets.

The IPCC noted the omission of this component in its sea level projections, because it was unable to reach a consensus on the magnitude of likely ice sheet disintegration. However, much of the media failed to note this caveat in the IPCC report.

Earth's history reveals many cases when sea level rose several meters per century, in response to forcings much weaker than present human-made climate forcings. Iceberg discharge from Greenland and West Antarctica has recently accelerated. It is difficult to say how fast ice sheet disintegration will proceed, but this issue provides strong incentive for policy makers to slow down the human-made experiment with our planet.

Knowledge of climate sensitivity has improved markedly base on improving paleoclimate data. The information on climate sensitivity, combined with knowledge of how sea level responded to past global warming, has increased concern that we could will to our children a situation in which future sea level change is out of their control."

Hansen commented that "the planetary energy imbalance is the single most critical metric for the state of the Earth's climate. Ocean heat storage is the largest item in this imbalance; it needs to be measured more accurately, present problems being incomplete coverage of data in depth and latitude, and poor inter-calibration among different instruments. The other essential measurement for tracking the energy imbalance is continued precise monitoring of the ice sheets via gravity satellite measurements."

According to Hansen, sea level is now increasing at a rate of about 3.5 centimeters per century, with thermal expansion of the ocean, melting of alpine glaciers, and the Greenland and West Antarctic ice sheets all contributing to this sea level rise. That is double the rate of 20 years ago, and that in turn was faster than the rate a century earlier. Previously, the sea level had been quite stable for the past several millennia.

Chapter Three — Climate Change

> He stood leaning on the gritty concrete rail. Perhaps in the world's destruction it would be possible at last to see how it was made. Oceans, mountains. The ponderous counter spectacle of things ceasing to be. The sweeping waste, hydroptic and coldly secular. The silence.
> — Cormac McCarthy, in *The Road*

> Human beings can't bear this duality either in the world or in themselves. They can't bear failing the world by their very existence, or the world failing them. They've sown disorder everywhere, and in wishing to perfect the world, they end up failing themselves. Self-hatred fuels the whole technological effort to make the world over anew. It's on this failing of existence that all religions thrive. You have to pay. In the past, it was God who took the reprisals; now we do it. We have undertaken to inflict the worst on ourselves, and to engineer our disappearance in an extremely complex and sophisticated way, in order to restore the world to the pure state it was in before we were in it.
> — The late philosopher Jean Baudrillard

Despite the best efforts and substantial cash of the global warming deniers, the science is clear. Only Luddites or those with a vested economic interest could dispute these findings. In fact, computer models — assailed by the skeptics — *have* been in error. Change is happening much more rapidly than the models predicted. Valuable time has been lost. And nature is about to wreak her revenge in a way that will become evident by 2030. Humans may not disappear but billions of lives will be radically changed across the globe. And the United States will not emerge unscathed.

Climate science is complicated. But the bare-bones facts are simple: there is a single railroad track with the engine of high-growth global capitalism flying toward another engine charging in the opposite direction — global climate change. High-growth capitalism will win in the short term but nature will ultimately prevail. It's that simple even if the science and politics are complex. And science is being sliced and diced in the political meat grinder. Scientists have done a poor job conveying the urgency of the problems and presenting them in ways that the public can understand. And self-serving politicians interested in short-term gain and corporate patronage have stalled any serious action in the past and are likely to do so in the future.

New revelations are reported blithely. Consider this brief wrap-up in the January 2008 issue of *Harper's* magazine: "A new climate-change model, which takes into account other models' failures to anticipate the rapidity of sea-ice disappearance, predicted that the Arctic Ocean will have ice-free summers within five years; a topographical survey found that rising sea levels will contaminate up to 40 percent more freshwater than previously estimated; and a British study found that natural disasters have quadrupled in the past two decades."

Sleep well, Mr. and Mrs. America, as you try to maintain your falling living standards, avoid financial disaster, bail out Wall Street, and survive for one more day, a month, or a year. Don't examine things too closely or that Lunesta won't kick in as well. And try not to think about the legacy you will leave for Junior or little Melissa. Sleep well. Too much knowledge is a dangerous thing and often uncomfortable — and the higher taxes needed to combat climate change are now politically out of the question.

But people like Pablo Rafael Gonzalez are probably not sleeping well at all. Gonzalez has examined the total picture — not just a slice of it. As one domino falls, he understands, the others will follow. Researcher must consider the science, the economics, and the proclivities of *Homo sapiens* themselves when faced with crisis. Connect the dots. The prognosis is frightening. Are his views radical? If so, he has been joined by unlikely allies like the U.S. insurance industry and the Pentagon.

"In the end," Gonzalez writes in his book *Running Out: How Global Shortages Change the Economic Paradigm*: "after a long period of research, the strongest impression that remained with me was that the crisis of a permanent oil shortage was equaled by the crisis caused by the pollution that petroleum use generates. We need the energy derived from fossil fuels; no adequate substitution for them has been discovered. But if we continue using energy the way we do, the damage that is caused to the entire ecosystem is of such magnitude that in a very short time there will be no life left on Earth."[1]

The hypothesis that oil is running out is documented by official statistics on oil production and oil reserves. Reserves are declining measurably in various parts of the world, in a way that has no parallel in history. In the past, mankind

---

1 Gonzalez, ibid., p. 4.

chose to replace wood and coal with other fuels because they were more efficient or more convenient, not because the old fuel was running out all over the world. We have already picked the produce on the low-hanging branches. Extraction will become much more difficult, expensive, and energy-intensive.

Saudi Arabia, for instance, has taken great pains recently to assure the world that the kingdom will reliably continue to supply the world with the bulk of its oil needs well into the future. On December 7, 2008, a *60 Minutes* segment focused on the health of Saudi oil production. The state oil industry granted unprecedented access to the producers. In fact, Saudi Aramco has staged something of a marketing blitz recently which is unusual for the ordinarily tight-lipped oil producer. That begs the question: *Why?*

The answer may very well be found in a compelling book written by Matthew R. Simmons entitled *Twilight in the Desert: The Coming Saudi Oil Shock and the World Economy*. Simmons contends that Saudi Arabia has been less than honest about its oil reserves.[1]

Simmons cannot be easily dismissed. He is chairman and CEO of Simmons & Company International, a Houston-based investment bank that specializes in the energy industry. Simmons also serves on the boards of Brown-Forman Corporation and The Atlantic Council of the United States. He is also a member of the National Petroleum Council and the Council on Foreign Relations. He has an MBA from Harvard and has had three-plus decades of insider experience in the oil business. For the book, Simmons reviewed more than 200 independently produced reports about Saudi petroleum resources and production operations. He knows the business.

Most analysts already believe that countries routinely overestimate oil reserves for various economic and political reasons. Simmons writes, "Following nationalization, Saudi Aramco built on [its] foundation to become the world's largest integrated oil company, wielding leading-edge technology under best-in-class management. The company's record over several decades has been one of *responsible stewardship* of the nation's petroleum resources (although it has not been particularly successful in augmenting them) and *paternalistic care* for world oil markets."

But Simmons contends that their future is less assured.

> The laws of nature began to assert themselves in the Saudi oilfields during the 1970s, threatening to end the easy-oil era of free-flowing bounty. Reality came flooding into the giant and super-giant fields, literally in the form of water encroachment, as well as in more frequent, incremental pressure declines. These problems increased and became more severe during the 1980s and 1990s, requiring the use of more sophisticated technology and raising Saudi Arabia's costs of production. It

1 Matthew R. Simmons, *Twilight in the Desert: The Coming Saudi Oil Shock and the World Economy* (Hoboken, N. J.: John Wiley & Sons, Inc.), 2005, p. 100.

remains to be seen if the technology will deliver all the desired results without any unexpected consequences.

Simmons pored over technical papers from as far back as the early 1960s written by various professionals and reached the conclusion that the "problems they describe are real, and the picture that emerges from them undermines the optimistic but unsubstantiated claims of Saudi officialdom....Taken together and properly interpreted, they show us that the geological phenomena and natural driving forces that created the Saudi oil miracle are conspiring now in normal and predictable ways to bring it to its conclusion, in a timeframe potentially far shorter than officialdom would have us believe."[1]

Saudi Arabia is also experiencing domestic problems — an exploding population, water problems, and decreasing per capita funds for its welfare state.[2]

> The descriptions of these deep ... gas wells in the technical literature bear a remarkable similarity to descriptions of the deep gas wells now being drilled in south Texas, which have also proven to be difficult to drill, expensive to produce, and quick to decline following peak production. The deep gas resources in Texas are being exploited only because more easily drilled shallower gas accumulations have largely been depleted. It's the only gas play left for south Texas oilmen.[3]

> The gas challenge is now a serious threat to Saudi Arabia's long-term well-being. The oil challenge is mostly a *monetary* threat for the kingdom, although it is also a major concern for the rest of the oil-consuming world. Saudi Arabia's need for a substantial increase in natural gas supply is fundamental, a more compelling concern for the kingdom than flattening or even declining oil production. The kingdom's natural gas needs are all about creating more kilowatts and more potable water. Without abundant additions to natural gas supplies, it will be extraordinarily difficult, and perhaps impossible, for Saudi Arabia to fill these urgent *social* needs. The challenge of Saudi Arabia's power and desalination needs is poorly understood outside the kingdom. Little information has been published on desalination globally, let alone the need for massive desalination expansion in Saudi Arabia.[4]

> On the face of it, [the technical reports] should make us skeptical, at the very least, of the two principal claims endlessly reiterated by the Saudi Arabian petroleum authorities: 1. The proven oil reserves remaining in the aging giants and an array of lesser fields amount to something over 260 billion barrels. 2. The desert kingdom will be able to raise oil production to the level of 15, 20, or even 25 million barrels per day demanded by long-term energy forecasts. Saying it does not make it so. *Repetition is not persuasion.*[5]

---

1 Ibid., p. 100.
2 Ibid., pp. 247-248.
3 Ibid., p. 253.
4 Ibid., p. 260.
5 Ibid., pp. 261-262.

Simmons believes that crude oil peaked globally in 2005.

There is no way they can replace even a portion of hydrocarbon energy," Simmons said in a radio interview with Wilderness.com. When asked about a solution, Simmons said, "I don't think there is one.... The solution is to pray. Pray for mild weather and a mild winter.

Simmons told *Fortune* magazine that he expects oil prices to rise to $500 a barrel in the coming years.[1] He also told the magazine that "we don't have the ability to come to grips with a crisis until it hits us in the face." He noted that world oil production today is at its peak at 86 million barrels. According to the International Energy Agency in Paris, demand will be 115 million barrels by 2030.

Simmons was also skeptical about the future of oil shale which has been touted as a new energy source — especially the fields in Alberta, Canada.

"It's Buck Rogers stuff," Simmons commented. "It just can't work."

Ethanol?

"It's a joke. The numbers don't add up."

Simmons said that the world needs "big-time conservation, not feel-good conservation."

The price of oil has never reflected its true scarcity. In only six years, between 1974 and 1980, it leapt by 3,200 percent from $1.25 per barrel to $40 per barrel. In 2005, it jumped dramatically again. We can anticipate worse shocks in the future when it becomes clear to everyone that the oil supply is being depleted very quickly. While the price of oil jumped to over $140 a barrel in 2008 only to crashed to $40 during the recent global economic downturn, prices will inevitably rebound once the worst of the crisis has passed.

That oil depletion is real cannot seriously be questioned. Even many "renewable" natural resources that should not have disappeared have, nevertheless, done just that. If even renewable resources are being used up, how much more readily will the non-renewable resources disappear? The extinction of animal and plant species and deforestation indexes all point in the same direction. Shrinking lakes and the decrease of river flows and greater pollution have caused serious problems for the supply of drinking water, irrigation, and for hydroelectricity generation in many regions.

Subsequently, a correlation is made among population, food production worldwide, water and energy, to demonstrate — with official figures — the observable deterioration in the supply of these indispensable commodities. This is correlated with the effects of pollution.

"Man can develop new forms of capital, just as he has done up to now," Pablo Rafael Gonzalez writes. "Money is his invention and machines are, too. But man cannot replace nature in its creative role as supplier. If, due to overexploitation or the effects of pollution, natural resources begin to dry up, the complete structure

---

1   Brian O'Keefe, "Here Comes $500 Oil," *Fortune Magazine*, Sept. 22, 2008.

on which the modern world was built will suffer major changes. The abundance or the lack of natural resources does not depend on man's will. Man can reach — and has reached — high production levels in agriculture under favorable natural conditions. But if those conditions disappear, due for example to the greenhouse effect, then man will not be able to repeat those achievements. The same applies to other productive and social activities. What happens with natural resources defines, in important ways, human life."

If the current rules of consumption follow their course, in the year 2025 two out of three humans will live under stressful conditions because of lack of water.

"Desertification and drought affected 900 million people in a hundred countries in 2000 and it is considered that by 2025 this number will be doubled, and therefore 25 percent of the earth's surface will be degraded," according to the World Bank.

And the worst is still to come: a real energy crisis, because man is using up the resources of fossil energy, especially oil.[1]

The problem is complex. Pollution alone has reached a level that endangers many forms of life. How mankind can go forward is the greatest challenge facing political leaders and scientists alike. If, out of an excessive pragmatism, they only concentrate on the problems at hand and fail to provide for the future, humanity as a whole will be under serious threat.[2]

The United States population consumes more oil than Japan, China, Germany, Russia, South Korea, and India combined and is the biggest industrial oil consumer in the world. In recent years the developed world has been transferring to the Third World part of its industrial structure, because these countries impose fewer regulations that protect the environment (and the workers). The general structure of costs and wages is smaller than in the developed countries.[3]

The world economy has two major points of vulnerability: natural population growth and the new demand derived from the incorporation into the labor market of a large segment of the population of the Third World countries.

"The global crisis in natural resources will very soon be felt if the industrialized countries maintain the average rate of economic growth and if even a very small expansion takes place in the economy of the countries of the Third World. The environmental implications are enormous. If the present production pattern persists, with its intense use of the most polluting energy sources, environmental destruction will also accelerate," according to Gonzalez.

In the developed countries, economic growth is stimulated by advertising because most of the population has already met its basic needs. As has become very clear, consumerism is the force that moves the economy — some 70 percent of it

---

1 Gonzalez, ibid., p. 9.
2 Ibid., p. 9.
3 Ibid., p. 20.

in the U.S. Some people are used to purchasing new cars every year. They can't be bothered to have appliances repaired. The economic structure of the developed countries is built to favor intense consumption.[1] Americans have cut consumption out of necessity in 2009 but it is unclear if that trend will continue. Gas consumption has risen again as prices have fallen.

The American public has been taken on a wild ride over past few years as prices at the pump and prices on the New York Stock Exchange have increased rapidly, decreased rapidly, and are slowly creeping back up.

Tim Middleton, writing for MSN Money in February of 2009, said Americans can expect gas prices to soar again.[2]

> Prices of petroleum and natural gas have plunged from nearly twice what most observers regard as reasonable to nearly half.
>
> Energy stocks have fallen right along with them. And though some have started to rebound, they're still well below the levels they hit when gas was $4 a gallon.
>
> It's pretty much a sure bet that fuel prices will rise again and that energy companies will cash in again. Oil has fallen from more than $147 a barrel to below $41 today, and few, if any, analysts think that will last long term.
>
> "The secular bull market (in energy) is intact," says Tim Guinness, the lead manager of the Guinness Atkinson Global Energy Fund. "There will be a breathing space over the next three years ... where oil trades between $60 and $80. It's not going over $100 for three years, at least, but go above $100 it will."

Middleton said that one reason energy prices will rise is that the ambitious plans of President Obama and other world leaders to develop alternatives to fossil fuels will take decades to have an effect. Meanwhile, consumers of energy are proving more sensitive to price than to the recession. Oil consumption has been declining, but the pace of decline has slowed as prices have come down.

> Advocates of alternative sources of energy ... acknowledge theirs will be a lengthy quest," according to Middleton. "If the United States reaches the goal of having 1 million plug-in hybrid cars on the road by 2013, they will still account for only 0.4 percent of the total number of cars. Wind farms and, especially, solar power for electricity generation could garner a greater market share than that, but it will still be measured in low percentages for at least a generation.
>
> And China's policies in coming years could put as many as 600 million more cars on the roads — 20 times as many as the country has now and more than twice as many cars as are on U.S. roads today.
>
> That all adds up to a huge increase in demand for limited supply. And makes higher energy prices one of the safer bets in the market today.

---

1 Ibid., p. 21.
2 Tim Middleton, "Why Gas Prices Will Soar Again," *MSN Money*, Feb. 3, 2009.

Middleton goes on to say that the "global economy will recover, probably beginning in the second half of [2009]. Energy will literally fuel this recovery and nowhere more so than in the rapidly developing nations of Asia, notably China. The great bulk of their populations have not yet benefitted from globalization, but the benefits have rained down so close that they are keenly aware of them, envy them bitterly and are beginning to demand a share of them aggressively. Political leaders who don't deliver won't survive."

Middleton quoted Clare McKeen of *Barron's* as saying that oil prices are likely to rise because conservation efforts and alternative energies are slow to have an effect.

The world has enjoyed an abundance of natural resources and manpower, but has used them to lavish goods upon just 10 percent of the world population. When 90 percent of the population eked by on very little, the demands on nature were limited. But now, the 90 percent are more aware of the outside world, and they want their share. The economy will have to face a new paradigm. By the intense use of capital resources to boost production, we are quickly exhausting the supply of irreplaceable materials.

What will we do without freshwater? Rationing has already pitted residential areas against farmers in California and elsewhere. Efficient and affordable methods of purifying and desalinating water on a massive scale have yet to be found; neither have the attempts to find new sources of energy come up with a substitute for petroleum products. Science and technology must be directed to finding the most efficient ways to use the resources we have, but neither science nor technology can replace the vital resources.[1] The massive transportation sector in the U.S. relies on petroleum and there is no replacement in sight.

Using energy sources generates pollution, especially when it comes to fossil fuels such as oil, coal, and natural gas (which represented more than 90 percent of the world's energy consumption in 1999). As a consequence, pollution is inevitable, and directly linked to population growth. And because population constantly grows, the production of energy is also constantly growing. One common factor correlates natural resources, the economy, and the environment: the population of the world, which is growing every day, especially in the poorest regions.[2]

Oil supplies 40 percent of the world's energy needs. A real oil crisis would indeed beat any Hollywood disaster movie. It is a problem of multiple implications. Economic growth is linked to oil consumption growth. If growth diminishes, the economies won't be able to generate jobs for the new generations. A recession feedback loop would ensue that could not be broken without elevating oil consumption. This is the paradox: to maintain economic growth, we have to

---

1 Gonzalez, ibid., p. 28.
2 Ibid., p. 29.

consume more oil. But the more oil that is consumed, the sooner it will run out. In turn, as the gravity of the problem sinks in, prices will skyrocket until they reflect the value of petroleum's scarcity. Hyperinflation will cause world trade to contract. Economic growth will be a dream of the past.[1]

Journalism students are taught early that the public is generally not interested in what might happen far in the future, nor in what happens in faraway places. "But when that reality is recognized, there will be cataclysm," Gonzalez predicts.

Under these circumstances, he says the choice should be clear. The world will opt to go on as it has in the past, without worrying much about what might happen in the future. Oil consumption will continue to rise, and pollution, global warming, and the deterioration of the ozone layer will intensify. For now, there is no way to promote economic growth without consuming more oil and coal.

Gonzalez argues that the truth is not a popular commodity with the public or the media because it is unpleasant. Most people have little tolerance for it. "Everybody has his own fantasy and is happy believing in it. Politicians of every stripe, democratic or authoritarian, and economists, those who believe in the free market as well as socialists, each in his own way lives his own illusion and tries to express it to others," Gonzales writes. One of the common fantasies in the modern world — stimulated by many interested sectors — is the belief that Earth's resources will last forever. Theological beliefs have not helped either — that God, not man, is in charge.

"Most people, in most countries, take for granted that there is plenty of oil and therefore the topic does not generate any concern. But oil is already beginning to run out. That the Middle East (home to three-fourths of the proven oil reserves) has become a battleground where the most powerful nations skirmish shamelessly, pounding each other and everyone who gets in the way, reveals the ruthless conflict that must ensue when the word gets out that the oil is almost gone," Gonzalez writes.

"By 2004, the official studies of the International Union for Conservation of Nature and Natural Resources noted that 237 wild species had disappeared in the United States since observers began taking note. The country with the second greatest number of missing species for 2004 was French Polynesia, with 76 species reaching extinction. Mankind has sometimes claimed for himself a role as 'steward of the earth.' Any steward who lost this much of what was under his care would surely be fired, and it is not inconceivable that nature is about to get rid of this steward, as well," he adds.

The International Union for Conservation of Nature and Natural Resources also provides numbers on species in danger of extinction. In 2004, the United States headed the list with a total of 903 species at risk, followed by Australia

---

1 Ibid., p. 34.

with 562.[1] Brazil leads the field with 22,264 square kilometers deforested between 1990 and 2000. In second place is Indonesia with 13,124 kilometers.[2]

According to the United Nations Report Geo2000, between 1980 and 1990 Latin America lost 5.5 million hectares of tropical forests. Of the total deforestation worldwide, 25 percent of the destruction was in the tropical rainforest and 43 percent in subtropical forests. In Central America the dry forests were reduced drastically to only 4 percent of their original area.[3]

The World Bank considers that "nearly half of the world forests have been destroyed, most of them during the last forty years of the 20th century. Also, 40 percent of the Earth's surface is arid and vulnerable to degradation."[4]

China has increased its use of coal. Coal is one of the biggest contributors of pollutants implicated in the greenhouse effect. Given the magnitude of China's population and its economy, any change in how China uses natural resources has an effect on the rest of the world.[5] The communist regime in China has made a bargain with the population: keep us in power and we will give you a more prosperous life. That prosperity will not be possible without using coal, which China has in abundance. During the recent economic downturn, China has indicated that economic recovery will trump climate change problems.

Some scientists and conservationists were hoping that a global economic slump would at least slow the pace of carbon dioxide emissions — one bright spot in an otherwise bleak scenario. This does not seem to be the case.

According to a February 2009 Associated Press story headlined "Carbon emissions show rapid rise; Growth attributed to the use of coal as cheap energy source," humans are adding carbon to the atmosphere even faster than in the 1990s.[6]

Researchers said that carbon emissions have been growing at 3.5 percent per year since 2000, up sharply from 0.9 percent per year in the 1990s.

"It is now outside the entire envelope of possibilities" considered in the 2007 report of the International Panel on Climate Change, according to Christopher Field of the Carnegie Institution for Science.

The largest factor in this increase is the widespread adoption of coal as a cheap energy source, Field said at the annual meeting of the American Association for the Advancement of Science.

"Without aggressive attention, societies will continue to focus on the energy sources that are cheapest, and that means coal," he said, adding that past projec-

1 Ibid., p. 42.
2 Ibid., p. 47.
3 Ibid., p. 47.
4 Ibid., p. 48.
5 Ibid., p. 50.
6 "Carbon Emissions Show Rapid Rise, Growth Attributed to the Use of Coal as Cheap Energy Source," The Associated Press, Feb. 15, 2009.

tions for declines in the emissions of greenhouse gases were too optimistic. He said that no part of the world had a decline in emissions from 2000 to 2008.

Anny Cazenave of France's National Center for Space Studies also reported that improved satellite measurements show that sea levels are rising faster than had been expected.

The highly promoted efforts to curb carbon emissions through the use of bio-fuels may even backfire, other researchers said. Demand for biologically-based fuels had led to the growing of more corn in the United States, but that resulted in switching fields from soybeans to corn, explained Michael Coe of the Woods Hole Research Center. But there was no decline in the demand for soy, he said, and countries like Brazil increased their soy crops to make up for the deficit. In turn, Brazil created more soy fields by destroying tropical forests, which tend to soak up carbon dioxide. Instead the forests were burned, releasing the gases into the air. The increased emissions from Brazil swamp any declines recorded by the United States, Coe said.

The U.S. Energy Department recently reported that the amount of U.S. green-house gases flowing into the atmosphere, mainly carbon dioxide from burning fossil fuels, increased 1.4 percent in 2007 after a decline in 2006. The report said carbon dioxide rose 1.3 percent in 2007 as people used more coal, oil, and natural gas because of a colder winter and more electricity during a warmer summer, adding that half the country's electricity is generated by coal-burning power plants.

For scientists like James Hansen, coal is the primary issue. He has said that "coal is the single greatest threat to civilization and all life on our planet" and has stated that "trains carrying coal to power plants are death trains" and "[c]oal-fired power plants are factories of death."

And coal is a particularly acute problem — both environmentally and politically.

"Coal is not only the largest fossil fuel reservoir of carbon dioxide, it is the dirtiest fuel," Hansen writes in an article entitled *The Sword of Damocles.* "Coal is polluting the world's oceans and streams with mercury, arsenic, and other dangerous chemicals. The dirtiest trick that governments play on their citizens is the pretense that they are working on 'clean coal' or that they will build power plants that are 'capture ready' in case technology is ever developed to capture all pollutants."

Hansen, through bitter experience working within the George W. Bush Administration, has come to understand the political problems and the power of vested interests.

> The public, buffeted by day-to-day weather fluctuations and economic turmoil, has little time or training to analyze decadal changes. How can they be expected to evaluate and filter out advice emanating from

special economic interests? How can they distinguish top-notch science and pseudoscience?

China and the United States have huge reserves of cheap, dirty coal. There will be tremendous pressure to use those reserves. Using those reserves will greatly exceed global climate tipping points. Tipping points will amplify other feedbacks with effects that would be irreversible.

In a letter to British Prime Minister Gordon Brown, Hansen stated his case:

"The German and Australian governments pretend to be green. When I show German officials that fossil fuel reservoir sizes imply that the coal source must be cut off, they say they will tighten the 'carbon cap.' But a cap only slows the use of a fuel; it does not leave it in the ground. When I point out that their new coal plants require that they convince Russia to leave its oil in the ground, they are silent. The Australian government was elected on a platform of solving the climate problem, but then, with the help of industry, they set emission targets so high as to guarantee untold disasters for the young and the unborn. These governments are not green. They are black — coal black.

"On a per capita basis, the three countries most responsible for fossil fuel carbon dioxide in the air today are the United Kingdom, the United States, and Germany.... Politicians in Britain have asked me: why am I speaking to them — the United States must lead. But coal interests have great power in the United States — the essential moratorium and phase-out of coal likely requires a growing public demand and a political will yet to be demonstrated.

"The Prime Minister should not underestimate his potential to initiate a transformative change of direction. And he must not pretend to be ignorant of the consequences of continuing coal emission, or take refuge in a 'carbon cap' or some 'target' for future emission reductions."

Hansen is no fan of President Obama's "cap-and-trade" proposal which he refers to as "tax-and-trade," a cry that has been taken up by many Republicans for different reasons. But many political pundits believe even this moderate step will be politically untenable because of soaring public debt and an ailing economy. "The biggest problem with Cap and Tax is that it will not solve the problem," Hansen said. "The public will soon learn that it is a tax.... There is no way the Cap Tax can get us back to 350 parts per million CO_2. The only solution to the climate problem is to leave much of the fossil fuels in the ground. That requires a high enough carbon price that we move on to our energy future beyond fossil fuels."

In 2003, a cap-and-trade proposal was defeated in the Senate by a vote of 55-43. However, this proposal would have an upside for Wall Street — an estimated $2 trillion carbon trading market.

The Center for Public Integrity also expressed the need for caution if "cap-and-trade" was controlled by market forces.

"With the global economy in meltdown, and faith in Wall Street wizardry at a low — to say the least — it's perhaps an odd time for a push to put the fate of the planet into the hands of the market," wrote the center's Marianne Lavelle in a February 2009 article "Carbon as a Commodity."[1]

"But that's the solution for fighting global warming already in practice in Europe, and the one with the most traction in Congress and the White House," she writes.

> When President Obama, soon after the election, promised "a new chapter" in American leadership on climate change, he said, "That will start with a cap-and-trade system."

> In cap-and-trade, Uncle Sam would either hand out or sell tradable "permits" that would allow power plants and other businesses to emit a certain amount of carbon dioxide into the atmosphere, and no more. The idea is that the permits would become a valuable commodity, and companies that can cut emissions quickly can profit by selling their permits to companies that are having a hard time. It's a way of giving businesses flexibility, while creating incentives for innovators to figure out the lowest-cost solutions.

> The idea of creating a "carbon market" is based on the hugely successful 29-year-old program that curbed U.S. acid rain far more quickly and cheaply than industry anticipated. But the acid rain market deals with only one pollutant — sulfur dioxide — and one kind of polluter: the coal power plant. Carbon dioxide and the other greenhouse gases like methane are so much more prevalent that an effort to limit them would involve every sector of the economy.

Commodity Futures Trading Commission member Bart Chilton said in a recent speech that cap-and-trade is "the most important thing we have never done." He anticipates that within five years a carbon market would dwarf any of the markets his agency now regulates — from livestock to corn to oil and natural gas. "I can see carbon trading being a $2 trillion market. The largest commodity market in the world."[2]

Lavelle writes that the carbon market would probably be the most complicated.

> The idea is that companies could meet their carbon 'caps' not only by cutting their own smokestack emissions, but with the help of 'offsets.' That is, by making investments in a wide-ranging array of projects that reduce greenhouse gas some other way — solar or wind power plants, tree planting, or capturing methane on hog farms.

> Each potential offset is fraught with possible controversy: Is it real? Is it verifiable? To move money around in such a market would require project developers, financiers, verifiers, registries, and consultants — all

---

1 Marianne Lavelle, "Carbon as a Commodity," The Center for Public Integrity, Feb. 24, 2009.
2 Ibid.

of whom are now well aware of their stake in climate policy. Wall Street banks like Goldman Sachs and JP Morgan Chase, insurance companies like AIG and private equity firms had virtually no reps on Capitol Hill working on global warming policy in 2003; by last year, they had about 130 climate lobbyists.... About 20 additional lobbyists worked for firms and organizations wholly dedicated to carbon marketing last year.

But given the shaky state of the world's financial markets, cap-and-trade advocates face a new skepticism on Capitol Hill. Senate Energy Committee Chairman Jeff Bingaman has voiced concerns that Congress will set a seemingly aggressive cap on carbon emissions, and then enact a system that allows such generous offsets that no emission reductions will actually take place.

Lavelle notes that Dirk Forrister, managing director and lobbyist for Natsource, a top finance firm in the carbon market that is already active in Europe under the Kyoto treaty, argues that there are ways the United States could already be cutting its greenhouse gas emissions today — through renewable energy, better agriculture, and forestry practices. But Lavelle writes that "those things aren't happening — not on the scale needed — because no one sees the value of investing in them. In theory, there would be a potential return on such investments under cap-and-trade." But Forrister, who chaired the White House Task Force on Climate Change under President Clinton, told Lavelle that in the wake of the recent financial upheaval, cap-and-trade advocates will find themselves in the coming months making the case not only for a market — but for regulation. "There's a newfound concern about the role of markets in general," Forrister said. "and I think we have to be thoughtful about what forms of market oversight need to be in place."[1]

Much has been made of so-called Carbon Capture and Sequestration (CSS) technology. The theory is to capture carbon dioxide gas — the byproduct of burning coal — compress it into liquid, and inject it deep into layers of rock. The Environmental Protection Agency says more than 4 trillion gallons of industrial waste has been pumped underground during the past 30 years.

Some environmental groups strongly oppose carbon capture.

"There is no way that it makes any sense to take liquid wastes and inject them into cavities deep in the Earth," said Stephen Lester, science director at the Center for Health, Environment and Justice. "It's a process destined to create problems for future generations." He said plants should install treatment systems, use better methods to recycle wastes, and find ways to reduce the creation and use of toxic compounds in manufacturing. Companies prefer injection, he said, because it's cheaper than other options.[2]

---

1 Ibid.
2 Spencer Hunt, "Idea of Burying Carbon Dioxide Gains Attention; Millions Being Spent on Studies," *Columbus* (Ohio) *Dispatch*, Feb. 9, 2009.

President Obama has pledged that five commercial-scale coal plants will be constructed with CSS technology. However, energy analysts have estimated that these plants will increase the cost of energy from 30 to 60 percent.

And the one foray into developing a CCS coal plant was a dismal failure. Called the "Future Gen" project, which was partially funded by the Department of Energy, it ended when the Department of Energy cancelled its funding in January 2008 due to "rising costs" which were expected to reach $1.8 billion. In addition to cost overruns, the project fell behind schedule. The costs were high because Future Gen was competing with new conventional coal plants throughout the world.

The cost of a full-scale CSS plant is expected to be about $1.3 billion, according to *The Economist*.[1] The technology is not expected to mature until after 2030 and, at best, the first plant could be ready by 2015.

Contrica, a British utility, estimated that it would take 15 to 20 years to roll out CSS plants in large numbers. *The Economist* concluded in its March 2009 article that CSS technology "is mostly hot air."

And the coal industry has not been idle. The National Mining Association has been ramping up its lobbying efforts, according to Jim Snyder of *The Hill* published in September of 2007.[2] He notes that coal accounts for 52 percent of electricity produced in the U.S. He also wrote that increased funds from lobbyists are now being funneled to Democrats. The American Coalition for Clean Coal Electricity, a consortium of 48 mining firms, railroads, and power companies, spent $10.5 million on public relations during 2008.

According to the World Coal Institute, the world reserves of coal are more than 1 trillion U.S. tons which would last approximately 133 years at current consumption rates.[3]

In the decades since 1970, the production of food worldwide grew at a moderate rate but the world population has grown dramatically. In 1950, the world population was 2.5 billion; fifty-five years later, in 2005, the population was 6.5 billion people. At that rate of expansion, by 2044 Earth will have 13 billion inhabitants.[4] This is well beyond the Earth's carrying capacity.

The water supply is also jeopardized by man's predatory attitude. If the deterioration continues, the Amazon region — that is the biggest forest reserve that is still left in the world — will soon become a desert.[5]

The Jordan River has also been shrinking and this only intensifies conflict between Arabs and Jews.

---

1 "The Illusion of Clean Coal," *The Economist*, March 5, 2009.
2 Jim Snyder, *"Facing New Threats, Mining Industry Undergoes a Facelift,"* The Hill, Sept. 18, 2007.
3 *"Coal Statistics,"* World Coal Institute, London, England, 2007.
4 Gonzalez, ibid., p. 52.
5 Ibid., p. 57.

Populations that do not have enough water or food may find no other alternative than to fight each other for the little that does exist. Beyond ideological and religious differences, at bottom the Middle East crisis can be seen in the light of competition for vital natural resources. The Sahara Desert is inexorably expanding, forcing the North African population to emigrate — mostly toward Europe, which creates other sorts of religious, ethnic, and cultural problems.[1] As discussed above, conflicts among states that share the waters of the Colorado River in the U.S. are also growing.

The need for water is fundamental; but so is the need for fuel. There is no economic growth without a more intense use of energy and, in the modern world, energy is synonymous with petroleum. But the use of energy has a high price: the pollution and destruction of the planet. Man is the only species known to destroy its own habitat. Wild animals do not destroy their nests or their breeding grounds, or the area where they live and hunt. Man does.[2]

Gonzalez outlines the world's current oil reserves and oil production rates, and indicates where they are increasing — or, more often, decreasing, while consumption inexorably rises.

Great faith had been placed in future discoveries, when as-yet-uninvented technologies would enable us to keep on finding and accessing presumed reserves in more remote places, such as under the deep layer of salt in the Gulf of Mexico. Some such innovations have indeed been put in place, but production and consumption continue to be increased faster than new discoveries are made; we are simply reducing the total resources.[3] U.S. oil companies are putting great hope (and a lot of money) into extracting oil out of tar sands in Alberta, Canada. Squeezing oil out of sand is environmentally destructive, as Alberta residents are finding out, and the refining process is especially dirty.[4]

In recent years, there has been very little progress in discovering "new" oil layers. It begins to seem that there is not a lot of oil left to be found.[5]

At this rate, the next oil shock will be more shocking than anything we've seen — and brought on largely by man's heedlessness.

Energy is closely linked with national security and power in every country. There is every reason to think that the official figures on oil reserves are imprecise at best. Whether they are overestimated or underestimated would depend on how each country perceives its own interest. Estimating high tends to magnify the apparent supply and to head off price increases. Estimating low could encourage price hikes on the basis of a perceived potential shortage.

---

1 Ibid., pp. 60-61
2 Ibid., p. 66.
3 Ibid., p. 87.
4 Michael Brune and Kenny Bruno, "Palin's Pipeline From Hell," *The Huffington Post*, June 17, 2009.
5 Gonzalez, ibid., p. 91.

The world has never yet experienced a crisis from sheer lack of oil. The price shocks of 1973 and 1979 were caused by political decisions, as the Arab countries moved in reprisal for the help the Western world gave to Israel. A genuine crisis of supply has not yet been seen. "As long as oil has been abundant, the oil producers have retained a degree of sovereignty, but when the oil shortage is apparent, the gloves will come off and the competition between the various world powers will create a new political and military situation," Gonzalez argues.[1]

In spite of, or in addition to, all other possible factors that contributed to the first Gulf War and now the Iraq War, the urge to take control of some of the greatest oil reserves in the world, and the desire to put more Western military installations in the region, should not be underestimated. How many other conflicts have had similar objectives?

The world is facing an unprecedented problem. As the population continues growing, demand for oil and other resources goes up while pollution and climate change will demand reductions. "Urgent, concerted action is required," Gonzalez asserts. "But do we have the political will to make tough choices? The answer seems to be that we do not."

Constant economic growth and globalization cannot go on in an unlimited way forever because the Earth's resources are not unlimited. More efficient technologies and alternatives can go a long way to extending the time frame, but the end must come sooner or later. Resources run out. Technology cannot replace Mother Nature as a primary source; it can only improve the yields.[2]

But so far, this ship is still moving fast and in the wrong direction. There is no sign of a miracle on the horizon. The recent recession in the U.S. is expected to cost taxpayers at least $3 trillion — deflecting money and priorities away from any substantial action on climate change.

According to Alan Durning, author of the book *How Much Is Enough?*, only population growth rivals high consumption as a cause of ecological decline, and at least population growth is now viewed as a problem by many governments and citizens of the world.[3]

"We may be in a conundrum — a problem with no satisfactory solution," Durning writes. "Limiting the consumer lifestyle to those who have already attained it is not politically possible, morally defensible, or ecologically sufficient. And extending that lifestyle to all would simply hasten the ruin of the biosphere. The global environment cannot support 1.1 billion people living like American consumers, much less 6 billion people, or a future population of at least 8 billion. On the other hand, reducing consumption levels of the consumer society, and tem-

---

1 Ibid., p. 195.
2 Ibid., p. 216.
3 Alan Durning, *How Much is Enough? The Consumer Society and the Future of the Earth* (New York: W.W. Norton & Company, Inc., 1992).

pering material aspirations elsewhere, though morally acceptable, is a quixotic proposal. It bucks the trends of centuries. Yet it may be the only option.

"The spread of the consumer lifestyle around the world marks the most rapid and fundamental change in day-to-day existence the human species has ever experienced. Over a few generations, we have become car drivers, television watchers, mall shoppers, and throwaway buyers. The tragic irony of this momentous transition is that the historic rise of the consumer society has been quite effective in harming the environment, but not in providing people with a fulfilling life."

A growing number of oil industry executives are now endorsing an idea long deemed fringe: The world is approaching a practical limit to the number of barrels of crude oil that can be pumped every day, according to a November 2007 article in the *Wall Street Journal*.[1]

Some predict that, despite the world's fast-growing thirst for oil, producers could hit that ceiling by 2012. This rough limit — which two senior industry officials pegged at about 100 million barrels a day — is well short of global demand projections over the next few decades. Prices are expected to rise dramatically as a "bare-knuckled" competition for oil ensues.

The new adherents — who range from senior Western oil executives to current and former officials of the major world exporting countries — share the belief that a global production ceiling is coming because of restricted access to oil fields, spiraling costs, and increasingly complex oil-field geology.

On October 31, 2007, Christophe de Margerie, the chief executive of French oil company Total SA, jolted attendees at a London conference by openly labeling production forecasts of the International Energy Agency as unrealistic. The IEA projects production will grow to between 102.3 million and 120 million barrels a day by 2030. De Margerie said production by 2030 of even 100 million barrels a day will be "difficult."

This is "the view of those who like to speak clearly, honestly, and (are) not just trying to please people," he said. The French executive said many existing oil fields are being depleted at rates that will damage their geologic structures which will limit future output. Even some OPEC officials are forecasting limits.

Sadad Ibrahim Al Husseini, a former head of exploration and production at Saudi Arabia's national oil company, has also gone public with doubts. He said in London recently that he didn't believe there were enough engineers or equipment to ramp up production fast enough to keep up with the economy.

Fueling passenger cars accounts for more than one-fourth of world oil consumption, and manufacturing them takes additional energy. Pumping that fuel from the Earth endangers ecosystems wherever it is pursued. And as oil companies drain easy-to-reach fields, they increasingly drill offshore or in remote areas

---

1 Russell Gold and Ann Davis, *"Oil Executives Believe Production Limit Near,"* The Wall Street Journal, Nov. 19, 2007.

such as Alaskan tundra and Ecuadoran rainforest. Oil refining, meanwhile, ranks first among U.S. manufacturing industries in energy intensity and fourth in total toxic emissions.

Beyond the pollution cars cause and the resources they absorb, accommodating them has profound impacts on land. In the United States, roads, parking lots, and other areas devoted to the car occupy half of urban space. And nationwide, pavement covers an area larger than the state of Georgia.[1]

Air travel makes even driving seem gentle on the environment. Jets use 40 percent more fuel than cars to move each passenger a kilometer, and much air travel has come not at the expense of car trips but of train and bus trips, thereby substituting the most energy-intensive form of long-distance travel for the least. Although airplanes do not cause much air pollution on the ground where people might inhale it, they pollute voluminously at higher altitudes.[2]

Durning quotes philosopher Ivan Illich who wrote in 1977, "Industrial society has created an urban landscape that is unfit for people unless they devour each day their own weight in metals and fuels." Average Americans in the early 1990s consumed, either directly or indirectly, 52 kilograms of basic materials a day — 18 kilograms of petroleum and coal, 13 of other minerals, 12 of agricultural products, and 9 of forest products.[3]

The chemicals industry produced millions of tons of synthetic substances — more than 70,000 varieties — many of which have proved impossible to isolate from the natural environment.[4]

The mind-boggling fact is the amount of time that has been wasted discussing environmental concerns (a vacuous debate kept alive primarily by self-interested companies) rather than addressing the myriad problems. American biologist Rachel Carson published her seminal book *Silent Spring* with this warning:

> The history of life on Earth has been a history of interaction between living things and their surroundings. To a large extent, the physical form and the habits of the Earth's vegetation and its animal life have been molded by the environment. Considering the whole span of earthly time, the opposite effect, in which life actually modifies its surroundings, has been relatively slight. Only within the moment of time represented by the present century has one species — man — acquired significant power to alter the nature of his world.

> During the past quarter century this power has not only increased to one of disturbing magnitude but it has changed in character. The most alarming of all man's assaults upon the environment is the contamination of air, earth, rivers, and sea with dangerous and even lethal materials. This pollution is for the most part irrecoverable; the chain of evil it

---

1 Durning, ibid., p. 82.
2 Ibid., p. 85.
3 Ibid., p. 92.
4 Ibid., p. 92.

initiates not only in the world that must support life but in living tissues is for the most part irreversible.

Carson also pointed out another problem which now looms large:

> There is still very limited awareness of the nature of the threat. This is an era of specialists, each of whom sees his own problem and is unaware of or intolerant of the larger frame into which it fits. It is also an era dominated by industry, in which the right to make a dollar at whatever cost is seldom challenged. When the public protests, confronted with some obvious evidence of damaging results of pesticide applications, it is fed little tranquilizing pills of half truth. We urgently need an end to these false assurances, to the sugar coating of unpalatable facts. It is the public that is being asked to assume the risks that the insect controllers calculate. The public must decide whether it wishes to continue on the present road, and it can do so only when in full possession of the facts. In the words of Jean Rostand, "The obligation to endure gives us the right to know."

Although advertising's main ecological danger is its promotion of consumerism, according to Durning, it also uses up lots of paper. Ads pack the daily mail — 14 billion mail-order catalogs plus 38 billion other assorted ads clog the post office each year in the United States. And they fill periodicals: a typical American newspaper is 65 percent advertising, up from 40 percent half a century ago.[1]

Since 1989, marketers have been painting their products "green" in an attempt to tap citizen anger at corporate ecological transgressions. In 1990, for example, the oil company Texaco offered Americans "free" tree seedlings to plant for the good of the environment; to qualify, a customer had to buy eight or more gallons of gasoline. Unmentioned in the marketing literature was the fact that it takes a typical tree about four years to store as much carbon dioxide as is released in refining and burning eight gallons of fuel, and that most tree seedlings planted by amateurs die.[2]

At its worst, green consumerism is a palliative for the conscience of the consumer class, allowing us to continue business as usual while feeling good about ourselves.[3]

Global Survival is a compilation of essays edited by Ervin Laszlo and Peter Seidel in 2006.[4] In the introduction, the problem is stated thusly:

> Many multinational corporations, whose only legal responsibility is to their stockholders, are larger, richer, and more powerful than many national governments. These businesses, with staffs of public relation experts and access to open global markets, interact with the environment, governments, societies, and the lives of each one of us — often to

---

1 Ibid., p. 122.
2 Ibid., p. 124.
3 Ibid., p. 125.
4 Ervin Laszlo and Peter Seidel, editors, *Global Survival: The Challenge and Its Implications for Thinking and Acting* (New York: SelectBooks, Inc., 2006).

---

our detriment. We know a lot of things that have nothing to do with survival, and are ignorant of many things that are essential for it. This affects how we treat our environment.

Seidel argues that it's time to devote more time to survival research.

As we head into the future, new discoveries and inventions will appear that will help us in significant ways we do not yet know," he writes. "But trusting science, technology, and unrestricted free enterprise and free trade to solve our problems is unwise. They cannot create more living space and cropland for a population that will continue to grow, replenish depleted aquifers, nor resurrect extinct species with their learned behaviors.

John H. Herz writes: "Regarding the ecological threat, the slowly evolving nature of change makes these dangers less visible to us. Every child born into the world contributes to the population explosion; however, its cumulative effect is not noticeable at any one moment. The same applies to the overuse of water supplies and energy resources, overfishing, deforestation, the pollution of 'just one more little stream or lake,' and so on. While people may know about such phenomena and their effects, they rarely grasp their impact on the global ecology."[1]

David and Marcia Pimentel, in their essay World Population, Food, Natural Resources, and Survival, write:

Soil scientists ... estimate that approximately 75 billion tons per year of soil are lost from world agricultural lands. Then, too, in many developing countries soil erosion is intensifying because fuel wood is in short supply and people burn crop residues as a fuel. Removal of crop residues leave soils unprotected from the effects of wind and rainfall, thereby intensifying erosion.

Agriculturists know that the fertility of nutrient-poor soil can be improved by large inputs of fossil fuel-based fertilizers. This practice, however, increases dependency on the finite fossil fuels stores used in the production of fertilizers. Even with current fertilizer use, soil erosion remains a critical problem in current agricultural production.

Crops can be grown under artificial conditions using hydroponic techniques, but the costs in terms of energy expenditure and dollars is approximately 10 times that of conventional agriculture. Such systems are not sustainable for the future. ...

Taken together, developed nations annually consume about 70 percent of the fossil energy worldwide, while the developing nations, which have about 75 percent of the world population, use only 30 percent. The United States, with only 4 percent of the world's population, consumes about 24 percent of the world's fossil energy output.

---

1 John H. Herz, essay "On Human Survival: Reflections on Survival Research and Survival Policies," *Global Survival: Reflections on Survival Research and Survival Policies,* editors Ervin Laszlo and Peter Seidel (New York: SelectBooks, Inc., 2006), p. 12.

Some developing nations that have especially high rates of population growth are increasing fossil fuel use to augment their agricultural production of food and fiber. For example, in China since 1955 there has been a 100-fold increase in fossil energy use in agriculture for fertilizers, pesticides, and irrigation."[1]

In *The Social Psychology of Sustainability*, David G. Myers writes: "With world economic growth and population both destined to increase (even as birth rates fall), resource depletion and global warming now seem inevitable. Ergo, the need for more sustainable consumption has taken on 'urgency and global significance.' The simple, stubborn fact is that the Earth cannot indefinitely support our present consumption, much less the expected increase in consumption. For our species to survive and flourish, some things must change."[2]

In *Economics Weak and Strong*, Andy Bahn and John Gowdy write:

Biodiversity loss is the other long-term threat to human survival. Humans evolved in the web of biodiversity on planet Earth, and current human activity is shredding this fabric of life. Before humans appeared on the planet, the rate of species extinction was about one per million per year, about the same rate as new species came into existence. Today, human activity has upped the rate of extinction to about 1,000 per year. Most of this extinction is occurring in a few biodiversity hotspots in tropical rainforests and coastal areas. Compared to the cost of stabilizing greenhouse gas emissions, the estimated cost of protecting the world's biodiversity hotspots is miniscule....[3]

Richard D. Lamm writes in his essay *Governance Barriers to Sustainability*:

I would argue that for all our genius, we can't escape ecological limits. I think the greatest challenge of the international community is to modify, and in some cases reverse, mores and cultural attitudes that have worked well and under which we have prospered for hundreds of years. We can delay, but not totally avoid, the consequences of infinite human demands on a finite earth. A very fundamental New World has emerged — a set of circumstances which are as important as the industrial or agricultural revolution. It is to change the world of growth into the world of sustainability.

I think the future can be better planned for by confronting limits to the best of our ability and heeding the warning that infinite growth cannot take place in a finite world. The fact that we have been so successful in pushing back those limits does not dissuade me from believing that those limits are real. 'All modern day curves lead to disaster,' warns former French President Valerie Giscard d'Estaing. (Population, consumption, environmental destruction.) Human civilizations are presently living on the upper shoulders of some incredibly steep and unprecedented

1 David Pimentel and Marcia Pimental, ibid., pp. 34-39.

2 David G. Myers, essay, "The Social Psychology of Sustainability," *Global Survival*, editors Ervin Laszlo and Peter Seidel (New York: SelectBooks, Inc., 2006), p. 102.

3 Andy Bahn and John Gowdy, essay *Economics Weak and Strong*, Global Survival, editors Ervin Laszlo and Peter Seidel (New York: SelectBooks, Inc., 2006), p. 184.

geometric curves. They cannot continue indefinitely. No trees grow to the sky. I believe that we are surrounded with evidence that increasingly shows that something is fundamentally wrong with the growth paradigm. Our globe is warming, our forests are shrinking, our ice caps are melting, our coral is dying, our fisheries are depleting, our deserts are encroaching, our finite water under more and more demand. I suspect these to be the early warning signs of a world approaching its carrying capacity. We cannot merely call upon human ingenuity, science and technology to develop new solutions to these new challenges. Technical know-how is not wisdom: it is important, but it is not enough.... We must be more than technically proficient; we must instead change our mental map of the world, our culture, and our economy.[1]

Jared Diamond, a professor of geography and physiology at UCLA, like many others, sees the threat posed by mass migrations with flows of environmental refugees banging on the gates of Europe and North America.

He notes that China alone faces massive problems. Along with severe deforestation and loss of wetlands, China faces destruction and degradation of grasslands due to overgrazing, climate change, and mining and other types of development. China's grasslands of the Tibetan Plateau (the world's largest high-altitude plateau) are headwaters for major rivers of India, Pakistan, Bangladesh, Thailand, Laos, Cambodia, and Vietnam as well as of China. Grassland degradation has increased the frequency and severity of dust storms in eastern China, notably Beijing.[2]

About one-fifth of China's native species, including the Giant Panda, are now endangered, and many other distinctive rare ones (such as Chinese alligators and ginkgos) are already at risk of extinction.[3]

For a year or two after the gas shortages of the 1973 Gulf oil crisis, we Americans shied away from gas-guzzling cars, but then we forgot the experience and embraced SUVs, despite volumes of print spilled over the 1973 events. When the city of Tucson, Arizona, went through a severe drought in the 1950s, its alarmed citizens swore that they would manage their water better, but soon returned to their water-guzzling ways of building golf courses and watering their gardens.

Diamond said we feel reluctant to abandon a policy in which we have already invested heavily. A prime example would be the Iraq war. In addition, we tend not to stand up to big business. Diamond blames the public for not demanding change. "It is rare that our society has effectively held the mining industry responsible for damages," Diamond states, adding, "In brief, environmental practices of big businesses are shaped by a fundamental fact that for many of us offends our sense of justice. Depending on the circumstances, a business really may

1 Richard D. Lamm, essay *Governance Barriers to Sustainability*, Global Survival, editors Ervin Laszlo and Peter Seidel (New York: SelectBooks, Inc., 2006), pp. 208-209.
2 Diamond, ibid., pp. 336-337.
3 Ibid., p. 337.

maximize its profits, at least in the short term, by damaging the environment and hurting people. When government regulation is effective, and when the public is environmentally aware, environmentally clean big businesses may outcompete dirty ones, but the reverse is likely to be true if government regulation is ineffective and if the public doesn't care."

Publicly owned companies with shareholders are under obligations to those shareholders to maximize profits, provided they do so by legal means.[1]

In Montana, California, and many other dry climates, the disappearance of mountain snow packs will decrease the water available for domestic uses, and for irrigation that actually limits crop yields in those areas. The rise in global sea levels as a result of snow and ice melting poses dangers of flooding and coastal erosion for densely populated low-lying coastal plains and river deltas already barely above or even below sea level. The eastern coast of the United States is especially vulnerable.[2]

Global warming will also produce big secondary effects that are difficult to predict exactly in advance and that are likely to cause huge problems, such as further climate changes resulting from changes in ocean circulation resulting in turn from melting of the Arctic ice cap.[3]

Even if the human populations of the developing countries did not exist, it would be impossible for the developed countries alone to maintain its present course, because it is not in a steady state but is depleting its own resources as well as those imported from the developing world. It is depleting its own resources as well as those imported from the developing nations. "It is untenable politically for First World leaders to propose to their own citizens that they lower their living standards, as measured by lower resource consumption and waste production rates," Diamond states. "What will happen when it finally dawns on all those people in the Third World that current First World standards are unreachable for them, and that the First World refuses to abandon those standards for itself?"[4]

An increase in population, obviously, leads to more deforestation, more toxic chemicals, more demand for wild fish, and other problems. The energy problem is linked to other problems because use of fossil fuels for energy greatly contributes to greenhouse gases, the combating of soil fertility losses by using synthetic fertilizers requires energy [including petroleum] to make the fertilizers, fossil fuel scarcity increases the push for in nuclear energy which poses potentially the biggest "toxic" problem of all in case of an accident, and fossil fuel scarcity also makes it more expensive to solve our freshwater problems by using vast amounts of energy to desalinize ocean water. Depletion of fisheries and other wild food

---

1 Ibid., p. 483.
2 Ibid., p. 493.
3 Ibid., p. 493.
4 Ibid., p. 496.

sources puts more pressure on livestock, crops, and aquaculture to replace them, thereby leading to more topsoil losses and more unnatural nourishment from agriculture and aquaculture. Problems of deforestation, water shortage, and soil degradation in the developing countries lead to wars there and drive legal asylum seekers and illegal emigrants to the developed world.[1]

"Our world society is presently on a non-sustainable course, and any of our problems would suffice to limit our lifestyle within the next several decades," according to Diamond. "They are like time bombs with fuses of less than 50 years."

"[B]ecause we are rapidly advancing along this non-sustainable course, the world's environmental problems *will* get resolved, in one way or another, within the lifetimes of the children and young adults alive today," Diamond writes. "The only question is whether they will come resolved in pleasant ways of our own choice, or in unpleasant ways not of our choice, such as warfare, genocide, starvation, disease epidemics, and collapses of societies. While all of those grim phenomena have been endemic to humanity throughout our history, their frequency increases with environmental degradation, population pressure, and the resulting poverty and political instability.

"Rich people will merely be the last people to starve or die," Diamond attests.

According to *Climate Change: A Groundwork Guide* written by Shelley Tanaka and published in 2006, modern human lifestyles now depend on consuming vast amounts of energy — to keep billions of people sheltered and warm; to produce, refrigerate, preserve, and transport food; to carry humans and cargo from place to place; and to produce the many goods that fill people's daily lives. The world's energy use has nearly doubled in the past 30 years and is expected to increase 60 percent by 2020.[2]

If the effects of climate change — heat stress, floods, droughts, insect-borne diseases, water shortages, shifting food sources — become chronic, many, many people will be on the move. Those who have access to air-conditioned buildings, who can afford to repair their storm- or flood-damaged homes and buy clean water, will probably be able to stay where they are. But for the hundreds of millions in the world who are already struggling, living from one harvest to the next, in countries with limited infrastructure and unstable governments, the migration will be desperate and massive.

Where will these people go? A 2003 report commissioned by the Pentagon concluded that climate change could create a security risk for the United States as early as 2020. With its large area, diverse climates, technological resources and wealth, the U.S. would be better able to cope than most, but it may decide to defend its borders and coastline to keep out those fleeing their own countries.

---

1 Ibid., p. 497.
2 Shelley Tanaka, *Climate Change: A Groundwork Guide* (Toronto: Anansi Press, 2006), p. 32.

The same thing could happen in other parts of the world. Central Europe, for instance, would become crowded with people fleeing from a thawing Scandinavia in the north and a drought-prone Mediterranean in the south.

And all over the globe there would be a battle for the world's arable land, energy and, especially, fresh water.[1]

If we continue to increase greenhouse gas emissions at the present rate, atmospheric carbon dioxide levels will double over the next 50 to 100 years. And if emissions increase even faster, the news will be much worse.[2]

China has a population of over 1.3 billion people, four times more than the United States. Although almost half the population still lives below the poverty line, in the past 20 years the average income in China has tripled and so has the country's energy use. The country is producing and buying more goods, and it is now the second-largest economy after the U.S. China now has 24 million cars on the road, and the country is making and selling a quarter of a million new cars each month. If China had as many cars per capita as the United States (135 million in 2002), there would be 600 million cars in the country — more than all the cars in the world today.

At a time when some countries are trying to shut down their coal-fired power plants (because coal is the most polluting fuel for carbon dioxide), China (which has vast supplies of coal) is planning to build more than 500 coal-fired plants by 2030. China is now the biggest consumer of coal and the second-largest user of oil after the U.S., and many experts say that with its urgent and massive energy needs (expected to double by 2020), it will overtake the U.S. as the largest producer of greenhouses gases by 2025.[3]

But the idea of reducing greenhouse gas emissions is, for many, simply unacceptable. Powerful forces — governments and corporations — actively campaign against climate change action by trying to dispute the science and by playing on the fears of people who don't want to put their good life at risk. The biggest and most profitable company in the world, ExxonMobil, refuses to invest in clean technologies and instead pressures governments to resist action on global warming. Governments in industrialized nations want to avoid the hard political decisions required to reduce carbon dioxide emissions. This is especially true during times of recession. Corporations fear that reducing the use of fossil fuels will mean reduced profits. Meanwhile, people in the developing world want their own economies to grow, which means using more energy.

It's true that people are gradually using more energy from renewable energy sources, but so far these industries are far too small to fill the world's needs, and each one has its drawbacks. Wind and solar power cannot be stored and must be

---

1 Ibid., p. 62.
2 Ibid., p. 67.
3 Ibid., p. 85.

backed up by other sources. And the initial set-up costs are expensive. It takes a lot of money and energy, for example, to make the photovoltaic panels needed to produce solar electric power. The International Energy Agency predicts that by 2030 renewables (excluding hydro) will contribute only 4 percent of the world's energy needs. Even nuclear power cannot fill the gap.[1]

"As Jared Diamond has pointed out, the world's environmental trouble spots are also the world's political trouble spots; the bigger the gap between rich and poor, the more politically unstable societies become," Tanaka observes.

By the mid-1970s, conservative economic and ideological interests had joined forces to combat what they saw as mindless eco-radicalism. Establishing conservative think tanks and media outlets, they propagated sophisticated intellectual arguments and expert public relation campaigns against government regulation for any purpose whatsoever. On global warming, it was naturally the fossil-fuel industries that took the lead. This resulted in elaborate studies or punchy advertisements aimed at persuading the public that there was nothing to worry about.

Scientists could themselves be co-opted so long as research grants were kept flowing by repeating the mantra: "More money should be spent on research."

Under Ronald Reagan, the EPA released a report on the greenhouse effect. The agency predicted a big rise within a few decades with potentially "catastrophic" consequences. Administration officials criticized the EPA report as alarmist.

Conservatives and skeptics might have been expected to oppose the creation of a prestigious body to study climate change. But they distrusted still more the system of international panels of independent scientists that had been driving the issue. The key word here is "international." Conservatives seem to believe that the rest of the world is out to get America.

In 1988, environmental organizations carried on with sporadic lobbying and advertising efforts to warn about global warming and argue for restrictions on emissions. They were opposed, and greatly outspent, by industries that produced or relied on fossil fuels. These industries channeled considerable sums of money to individual scientists and small conservative institutions and publications that denied any need to act against climate change.

Under Bill Clinton, conservatives retained a dominant influence in Washington politics. These conservatives scoffed at the research and had deep suspicions about the U.N.

"Politicians saw little to gain by stirring up (the global warming issue). Even Al Gore mentioned global warming only briefly during his 2000 run for the presidency," according to Spencer R. Weart, the author of *The Discovery of Global Warming*.

---

1 Ibid., p. 86.

Global warming certainly did not get the press that nuclear war threats did in the 1950s.

Weart recommends removing the huge government subsidies for fossil fuels and raising the tax on gasoline by a few dollars a gallon.

"Waiting for a sure answer (on global warming) would mean waiting forever," he writes.[1]

David Keys wrote in his 1999 book Catastrophe: An Investigation into the Origins of the Modern World:

> Climate has the potential to change history — not just on a short-term basis but in the long term as well.... Global warming (due to increased atmospheric pollution), sunspot activity, meteor or comet impacts, periodic small changes in the shape of the Earth's orbit and minor changes in the Earth's axis of rotation are all capable of triggering dramatic changes in climate — and human history.

An alarming study published in October 2007 found that global warming signals are stronger, and happening sooner than expected, due to increased human emissions of carbon dioxide and an Earth less able to absorb them. Carbon dioxide emissions were 35 percent higher in 2006 than in 1990, a much faster growth rate than anticipated, researchers reported in the peer-reviewed Proceedings of the National Academy of Sciences. Increased industrial use of fossil fuels coupled with a decline in the ability of land and oceans to absorb CO_2 were listed as causes of the increase. The changes "characterize a carbon cycle that is generating stronger-than-expected and sooner-than-expected climate forcing," the researchers wrote.[2]

On land, where plant growth is the major mechanism for soaking up CO_2, droughts have curbed that ability, they stated.

Finding ways to reduce greenhouse gas emissions has been a political problem because of potential effects on the economy.

According to the new study, carbon released from burning fossil fuel and making cement rose from 7 billion metric tons per year in 2000 to 8.4 billion metric tons in 2006. A metric ton is 2,205 pounds. The growth rate increased from 1.3 percent per year in 1990-1999 to 3.3 percent per year in 2000-2006, according to researchers. The research was supported by Australian, European, and other international agencies.

Has the public become convinced of the growing threat of climate change? Hardly.

---

1 Spencer R. Weart, *The Discovery of Global Warming* (Cambridge, Mass.: Harvard University Press, 2003).

2 *"Contributions to accelerating atmospheric CO_2 growth from economic activity, carbon intensity, and efficiency of natural sinks,"* Proceedings of the National Academy of Sciences Journal, vol. 104, issue 47, Oct. 22-26, 2007.

The Pew Center released new polling data in early 2007. The report showed that while 77 percent of people believe the Earth is warming, only 47 percent believe there is solid evidence that humans are responsible. A more recent Pew poll in 2009 shows that global warming came in dead last out of 20 political priorities in the U.S.[1]

The conservative, pro-business lobbies have famously followed the strategy laid out by Republican pollster Frank Luntz: "The scientific debate remains open. Voters believe that there is no consensus about global warming within the scientific community. Should the public come to believe that the scientific issues are settled, their views about global warming will change accordingly. Therefore, you need to continue to make the lack of scientific certainty a primary issue in the debate. The scientific debate is closing [against us] but not yet closed. There is still a window of opportunity to challenge the science."

Luntz amended his views in 2006 and said he now accepts that humans are affecting the climate based on the scientific evidence.[2]

Some political scientists believe many people will reject human-induced global warming — no matter how strong the evidence. The corporations have been very effective in maintaining confusion over the science of climate change.

The catastrophic fires that swept across Southern California in October 2007 are consistent with what climate change models have been predicting for years, according to *Science Daily*. Experts said they could be just a prelude to many more such events in the future — as vegetation grows heavier than usual and then ignites during prolonged periods of drought.[3]

"In the future, catastrophic fires such as those ... in California may simply be a normal part of the landscape," according to Ronald Neilson, a professor at Oregon State University and bioclimatologist with the USDA Forest Service.

Droughts or heat waves, researchers said in 2002, would then lead to levels of wildfires larger than that observed since European settlement. The projections were based on various "general circulation" models which showed both global warming and precipitation increases during the 21st century.

A 2006 study on global warming's impacts for Washington found the state already is fighting more forest fires per year. It predicted the state's direct costs would rise 50 percent by the 2020s. That doesn't include the loss of timber. Philip Mote, the state climatologist, has said insurance companies are already paying close attention to the Western risks.[4]

1 Matthew C. Nisbet, "*Pew: Global Warming Dead Last Among Public Priorities,*" ScienceBlogs, Jan. 22, 2009.
2 Faiz Shakir, "*Luntz Converts on Global Warming, Distances Himself From Bush,*" Think Progress, June 27, 2006.
3 "*Massive California Fires Consistent With Climate Change, Experts Say,*" Science Daily, adapted from materials provided by Oregon State University, Oct. 24, 2007.
4 "*Climate Change: Age of mega-fires,*" Seattle Post-Intelligencer, Oct. 24, 2007.

Forest fires in the Western United States have occurred more frequently, burned longer, and covered more acres since 1987 — and global warming is a big part of the underlying cause — according to a research paper published in July 2006 by the journal *Science*.

Researchers at the Scripps Institution of Oceanography and the University of Arizona found four times as many large wildfires occurred in Western forests between 1987 and 2003 compared to the previous 16 years. The more recent fires burned 6.5 times more land, the average duration of the fires increased from 7.8 to 37 days, and the overall fire season during those years grew by an average of 78 days.

Those changes corresponded to an average 1.5-degree rise in temperature throughout the American West during the same time period. According to the study, the first to link global warming to wildfires, the warmer temperatures due to climate change have led to longer, drier seasons, creating ideal conditions for forest fires.

> The real message of the paper is not as much about forest management," said Steven Running, a University of Montana ecology professor and one of the study's peer reviewers, according to an article in the *Sacramento Bee*. "It's that this is yet another dimension of global warming's impact. To me, it's the equivalent of the hurricanes on the Gulf Coast. This is our hurricane.[1]

Nationwide, wildfires burned an average of 3.6 million acres in the 1990s, but that number shot up to a record 8.4 million acres in 2000. That remained a record until 2005, when 8.6 million acres burned, according to the National Interagency Fire Center.

Federal agencies now spend more than $1 billion a year fighting wildfires, according to the study.

Like hurricanes along the Eastern Seaboard, forest fires have become a visible symbol for many in the West of the possible effects of climate change, according to Running. "Only a hurricane is over in a day or two; these forest fires can last for months."[2]

Globally, reports are becoming bleaker as the window of opportunity to halt or reverse climate change begins to close. *Scientific American* ran a story in October 2007 under the ominous headline "The World Is Not Enough for Humans." The article was prompted by the publication of the U.N.'s Environmental Program's Global Environmental Outlook.[3]

Since 1987 annual emissions of carbon dioxide have risen by a third, global fishing yields have declined by 10.6 million metric tons, and the amount of land

---

1 *"Global Warming Linked to Increase in U.S. Forest Fires,"* About.com: Environmental Issues, July 2006.

2 Dennis O'Brien, *"Climate Change Link Seen in Surge of Western Blazes,"* Baltimore Sun, July 7, 2006.

3 *"The World Is Not Enough for Humans: Humanity's environmental impact has reached an unprecedented scope, and it's getting worse,"* Scientific American, Oct. 26, 2007.

required to sustain humanity has swelled to more than 54 acres per person. Yet, Earth can provide only roughly 39 acres for every living person today. "There are no major issues," the report's authors write of the period since their first report in 1987, "for which the foreseeable trends are favorable."

And humanity's impact continues to grow. For example:

Biodiversity — The planet is in the grips of the sixth great extinction in its 4.5-billion-year history, this one largely manmade. Species are becoming extinct 100 times faster than the average rate in the fossil record. More than 30 percent of amphibians, 12 percent of birds and 23 percent of our own class, mammals, are threatened.

Climate — Average temperatures have climbed 1.4 degrees Fahrenheit over the past century and could increase as much as 8.1 degrees over the next century unless "drastic" steps are taken to reduce greenhouse gas emissions from, primarily, burning fossil fuels. Developed countries will need to reduce this globe-warming pollution by 60 to 80 percent by mid-century to stave off dire consequences, the report warns. "Fundamental changes in social and economic structures, including lifestyle changes, are crucial if rapid progress is to be achieved."

Food — The amount of food grown per acre has reached one metric ton, but such increasing intensity is also driving rapid desertification of formerly arable land as well as reliance on chemical pesticides and fertilizers. Four billion out of the world's 6.5 billion people could not get enough food to eat without such fertilization. Continuing population growth paired with a shift toward eating more meat leads the UNEP to predict that food demand may more than triple.

Water — One in 10 of the world's major rivers, including the Colorado and the Rio Grande in the U.S., fail to reach the sea for at least part of the year, due to demand for water. And that demand is rising; by 2025, the report predicts, demand for fresh water will rise by 50 percent in the developing world and 18 percent in industrialized countries. At the same time, human activity is polluting existing fresh waters with everything from fertilizer runoff to pharmaceuticals and climate change is shrinking the glaciers that provide drinking water for nearly one-third of humanity. "The escalating burden of water demand," the report says, "will become intolerable in water-scarce countries."

The authors — 388 scientists reviewed by roughly 1,000 of their peers — view the report as "an urgent call for action" and decry the "woefully inadequate" global response to climate change. "The amount of resources needed to sustain [humanity] exceeds what is available," the report declares.

"The systematic destruction of the Earth's natural and nature-based resources has reached a point where the economic viability of economies is being challenged," Achim Steiner, UNEP's executive director, said in a statement. "The bill we hand our children may prove impossible to pay."

Two institutions not normally associated with the environmental movement have also sounded the warning alarms — the insurance industry and the Pentagon.

According to the *Christian Science Monitor*, insurance companies are beginning to address climate change issues for economic reasons. This is the world's second-largest industry.[1]

Travelers, the giant insurance firm, offers owners of hybrid cars in California a 10 percent discount. It also offers the discount in 41 other states and has cornered a large share of the market. Fireman's Fund cut premiums for "green" buildings that save energy and emit fewer greenhouse gases. When it pays off claims, it directs customers to environmentally friendly products to replace roofs, windows, and water heaters. Marsh, the largest insurance broker in the U.S., offers a program with Yale University to teach corporate board members about their fiduciary responsibility to manage exposure to climate change.

Insurance companies, who like to stay out of the limelight, are becoming leading business protagonists in the assault on global warming. Their clout is sizable. It has a direct link to most homeowners and businesses. It insures coal-fired power plants as well as wind farms, so it can influence the power industry's cost structure. With its financial muscle, the industry could help advance the use of new financial instruments designed to allow companies to trade greenhouse-gas emissions in the same way that commodities are bought and sold.

Some consumers are already noticing a negative effect of this insurance shift. In 2006, some 600,000 homeowners living in a zone that an insurer considers a high storm risk in an era of climate change have seen their policies cancelled or not renewed. This includes coastal areas stretching from Texas to New York. Currently, coastal properties are valued at $7.2 trillion.

One reason for this massive change in coverage is an ongoing shift in the way insurance companies view risk. Insurers are starting to change their risk-assessment models to reflect future climate-change scenarios instead of past weather patterns.

The industry is not driven just by an attempt to help the environment: It also, of course, wants to make money. Travelers, for instance, recognized hybrid drivers as preferred customers — middle-aged, very responsible, and financially stable.

Insurance companies, adept at managing risk, are also trying to educate their customers. Marsh and Yale plan to train 200 board directors to understand risks of climate change. Again, part of the motivation is money: Insurance companies provide liability insurance for board members.

---

1 Ron Scherer, "*New Combatant Against Global Warming: Insurance Industry,*" The Christian Science Monitor, Oct. 13, 2006.

A Pentagon document released in 2004 predicted that abrupt climate change could bring the planet to the edge of anarchy as countries develop a nuclear threat to defend and secure dwindling food, water, and energy supplies. The threat to global stability vastly eclipses that of terrorism, say experts who have examined the report.[1]

Climate change "should be elevated beyond a scientific debate to a U.S. national security concern," wrote the authors, Peter Schwartz, CIA consultant and former head of planning at Royal Dutch/Shell Group, and Doug Randall of the California-based Global Business Network.

An imminent scenario of catastrophic climate change is "plausible and would challenge United States national security in ways that should be considered immediately," they concluded.

Bob Watson, chief scientist for the World Bank, and former chair of the IPCC, said that the Pentagon's dire warnings could no longer be ignored. His remarks were aimed at former President George W. Bush.

"Can Bush ignore the Pentagon? It's going to be hard to blow off this sort of document. It's hugely embarrassing. After all, Bush's single highest priority is national defense. The Pentagon is no wacko, liberal group; generally speaking, it is conservative. If climate change is a threat to national security and the economy, then he has to act. There are two groups the Bush Administration tend to listen to — the oil lobby and the Pentagon," added Wilson.

Already, according to Randall and Schwartz, the planet is carrying a higher population than it can sustain. By 2020, "catastrophic" shortages of water and energy supply will become increasingly harder to overcome, plunging the planet into war. They warn that 8,200 years ago climatic conditions brought widespread crop failure, famine, disease, and mass migration of populations that could soon be repeated.

Randall told *The Observer* that the potential ramifications of rapid climate change would create global chaos. "This is depressing stuff," he said. "It is a national security threat that is unique because there is no enemy to point your guns at and we have no control over the threat."

Randall added that it was already possibly too late to prevent a disaster from happening. "We don't know exactly where we are in the process. It could start tomorrow and we would not know for another five years," he said. "The consequences for some nations of the climate change are unbelievable. It seems obvious that cutting the use of fossil fuels would be worthwhile."

Jeremy Symons, a former whistleblower at the EPA, said that suppression of the report for four months was a further example of the Bush White House trying to bury the threat of climate change.

---

1 Mark Townsend and Paul Harris, *"Now the Pentagon Tells Bush: Climate Change Will Destroy Us,"* The Observer [U.K.], Feb. 22, 2004.

Symons, who left the EPA in protest over political interference, said: "It is yet another example of why this government should stop burying its head in the sand on this issue."

Symons said the former Bush Administration's close links to high-powered energy and oil companies was vital in understanding why climate change was received skeptically in the Oval Office. "This administration is ignoring the evidence in order to placate a handful of large energy and oil companies," he added.

Each day the world gulps down 82 million barrels of oil — virtually the same amount that is produced. The United States Energy Information Agency projects consumption to increase to 103 million barrels per day in 2015, and 119 million barrels each day by 2025. That means global production must increase by 45 percent — about five times the maximum annual output available from Alberta's oil sands — just to keep pace with ordinary economic growth.

There's just one problem. No one can say with confidence where all that extra oil will come from. It has been almost 60 years since Shell Oil senior geologist M. King Hubbert asserted in the journal of the American Association for the Advancement of Science that the dominance of fossil fuel in the global energy mix is just a tiny "pip" in the course of human history.

Michael T. Klare, a columnist for *Foreign Policy in Focus*, wrote that projected energy use and a habitable planet are mutually exclusive:

> Because Americans show no significant inclination to reduce their consumption of fossil fuels — but rather are using more and more of them all the time — one can foresee no future reduction in U.S. emissions of greenhouse gases. According to the Department of Energy, the United States is projected to consume 35 percent more oil, coal, and gas combined in 2030 than in 2004; not surprisingly, the nation's emissions of carbon dioxide are expected to rise by approximately the same percentage over this period. If these predictions prove accurate, total U.S. carbon dioxide emissions in 2030 will reach a staggering 8.1 billion metric tons, of which 42 percent will be generated through the consumption of oil (most of it in automobiles, vans, trucks, and buses), 40 percent by the burning of coal (principally to produce electricity), and the remainder by the combustion of natural gas (mainly for home heating and electricity generation). No other activity in the United States will come even close in terms of generating greenhouse gas emissions.

> What is true of the United States is also true of other industrialized and industrializing nations, including China and India. Although a few may rely on nuclear power or energy renewables to a greater extent than the United States, all continue to consume fossil fuels and to emit large quantities of carbon dioxide, and so all are contributing to the acceleration of global climate change. According to the DOE, global emissions of carbon dioxide are projected to increase by a frightening 75 percent between 2003 and 2030, from 25 to 43.7 billion metric tons. People may talk about slowing the rate of climate change, but if the figures prove ac-

curate, the climate will be much hotter in coming decades and this will produce the most damaging effects predicted by the IPCC.

What this tells us is that the global warming problem cannot be separated from the energy problem. If the human community continues to consume more fossil fuels to generate more energy, it inevitably will increase the emission of carbon dioxide and so hasten the build-up of greenhouse gases in the atmosphere, thus causing irreversible climate change. Whatever we do on the margins to ameliorate this process — such as planting trees to absorb some of the carbon emissions or slowing the rate of deforestation — will have only negligible effect so long as the central problem of fossil-fuel consumption is left unchecked.[1]

Many political and business leaders wish to deny this fundamental reality. They may claim to accept the conclusions of the IPCC report. They will admit that vigorous action is needed to stem the build-up of greenhouse gases. But they will nevertheless seek to shield energy policy from fundamental change.

"Typical of this approach was a talk given by Rex W. Tillerson, the CEO of ExxonMobil, at a conference organized by Cambridge Energy Research Associates in February of 2007," Klare writes." As head of the world's largest publicly-traded energy firm, Tillerson receives special attention when he talks. That his predecessor Lee Raymond often disparaged the science of global warming lent his comments particular significance. Yes, Tillerson admitted, atmospheric carbon dioxide levels were increasing, and this contributed to the planet's gradual warming. But then, in language characteristic of the industry, he added, 'The scale advantages of oil and natural gas across a broad array of applications provide economic value unmatched by any alternative.' It would therefore be a terrible mistake, he added, to rush into the development of energy alternatives when the consequences of global warming are still not fully understood.

"The logic of this mode of thinking is inescapable. The continued production of fossil fuels to sustain our existing economic system is too important to allow the health of the planet to stand in its way. Buy into this mode of thought, and you can say goodbye to any hope of slowing — let alone reversing — the build-up of greenhouse gases in the atmosphere."

According to Dr. Hansen of NASA, the warming since the beginning of the 20th century has been about 0.8 degrees Celsius, with three-quarters of that warming occurring in the past 30 years.

First, we must recognize that global mean temperature changes of even a few degrees or less can cause large climate impacts. Some of these impacts are associated with climate tipping points, in which large regional climate response happens rapidly as warming reaches critical levels. Already today's global temperature is near the level that will cause loss of all Arctic sea ice. Evidence suggests that we are also nearing the

1 Michael T. Klare, *"Global Warming: It's All About Energy,"* Foreign Policy In Focus," Feb. 15, 2007.

global temperature level that will cause the West Antarctic ice sheet and portions of the Greenland ice sheet to become unstable, with potential for very large sea level rise.

Second, we must recognize that there is more global warming 'in the pipeline' due to gases humans have already added to the air. The climate system has large thermal inertia, mainly due to the ocean, which averages 4 kilometers (about 2.5 miles) in depth. Because of the ocean's inertia, the planet warms up slowly in response to gases that humans are adding to the atmosphere. If atmospheric CO_2 and other gases stabilized at present amounts, the planet would still warm about 0.5 degrees Celsius (about 1 degree Fahrenheit) over the next century or two. In addition, there are more gases 'in the pipeline' due to existing infrastructure such as power plants and vehicles on the road. Even as the world begins to address global warming with improved technologies, the old infrastructure will add more gases, with still further warming on the order of another 1 degree Fahrenheit.

Third, eventual temperature increases will be much larger in critical high latitude regions than they are on average for the planet. High latitudes take longer to reach their equilibrium (long-term) response because the ocean mixes more deeply at high latitudes and because positive feedbacks increase the response time there. Amplification of high latitude warming is already beginning to show up in the Northern Hemisphere. Warming over land areas is larger than global mean warming, an expected consequence of the large ocean thermal inertia. Warming is larger at high latitudes than low latitudes, primarily because of the ice/snow albedo [reflected light] feedback. Warming is larger in the Northern Hemisphere than in the Southern Hemisphere, primarily because of greater ocean area in the Southern Hemisphere, and the fact that the entire Southern Ocean surface around Antarctica is cooled by deep mixing. Also human-caused depletion of stratospheric ozone, a greenhouse gas, has reduced warming over most of Antarctica. This ozone depletion and CO_2 increase have cooled the stratosphere, increased zonal winds around Antarctica, and thus warmed the Antarctic Peninsula while limiting warming of most of the Antarctic continent....

Until the past several years, warming has also been limited in Southern Greenland and the North Atlantic Ocean just southeast of Greenland, an expected effect of deep ocean mixing in that vicinity. However, recent warming on Greenland is approaching that of other land masses at similar latitudes in the Northern Hemisphere. On the long run, warming on the ice sheets is expected to be at least twice as large as global warming. Amplification of warming at high latitudes has practical consequences for the entire globe, especially via effects on ice sheets and sea level. High latitude amplification of warming is expected on theoretical grounds, it is found in climate models, and it is confirmed in paleo-climate [ancient climate] records.

Climate forcing is an imposed perturbation to the Earth's energy balance which would tend to alter the planet's temperature. An increase of greenhouse gases, which absorb terrestrial heat radiation and thus warm the Earth's surface, is also a positive forcing.

"The chief implication is that humans have taken control of global climate," Hansen stated. "This follows from figures that extend records of the principal greenhouse gases to the present. CO_2, CH_4 and N_2O are far outside their range of the past 800,000 years for which ice-core records of atmospheric composition are available."

But isn't the natural system driving the Earth toward colder climates?

If there were no humans on the planet, the long-term trend would be toward colder climate," according to Hansen. "However, the two principal mechanisms for attaining colder climate would be reduced greenhouse gas amounts and increased ice cover. The feeble natural processes that would push these mechanisms in that direction (toward less greenhouse gases and larger ice cover) are totally overwhelmed by human forcings. Greenhouse gas amounts are skyrocketing out of the normal range and ice is melting all over the planet. Humans now control global climate, for better or worse.

Another ice age cannot occur unless humans go extinct, or unless humans decide that they want an ice age. However, 'achieving' an ice age would be a huge task. In contrast, prevention of an ice age is a trivial task for humans, requiring only a 'thimbleful' of CFCs (chlorofluorocarbons), for example. The problem is rather the opposite; humans have already added enough greenhouse gases to the atmosphere to drive global temperature well above any level in the Holocene [a period which began about 11,700 years ago and continues to the present].

How much warmer will the Earth become because of the present level of greenhouse gases?

"That depends on how long we wait," Hansen said. "The Charney climate sensitivity (3 degrees Celsius global warming for doubled CO_2) does not include slow feedbacks, principally disintegration of ice sheets and poleward movement of vegetation as the planet warms.[1] When the long-lived greenhouse gases are changed arbitrarily, as humans are now doing, this change becomes the predominant forcing, and ice sheet and vegetation changes must be included as part of the response in determining long-term climate sensitivity."

What is the most useful geologic era to consider for gauging climate change?

The Cenozoic era, the past 65 million years, is particularly valuable for several reasons," according to Hansen. "First, we have the most complete and most accurate climate data for the most recent era. Second, climate changes in that era are large enough to include ice-free conditions. Third, we know that atmospheric greenhouse gases were the principal global forcing driving climate change in that era.

---

1 Jule Charney chaired a committee on anthropogenic global warming convened in 1979 by the National Academy of Sciences. The committee established the standard modern estimate of climate sensitivity.

For the past 50 million years and continuing today, regions of subduction of carbonate-rich ocean crust have been limited. Thus, while the oceans have been a strong sink for carbonate sediments, little carbonate is being subducted and returned to the atmosphere as CO_2. As a result, over the past 50 million years there has been a long-term decline of greenhouse gases and global temperature.

> In summary, there are many uncertainties about details of climate change during the Cenozoic era. Yet important conclusions emerge…. The dominant forcing that caused global cooling, from an ice-free planet to the present world with large ice sheets on two continents, was a decrease in atmospheric CO_2. Human-made rates of change of climate forcings, including CO_2, now dwarf the natural rates.

How do higher levels of global warming relate to dangerous climate change?

> Regional climate disruptions also deserve attention," Hansen points out. "Global warming intensifies the extremes of the hydrologic cycle. On the one hand, it increases the intensity of heavy rain and floods, as well as the maximum intensity of storms driven by latent heat, including thunderstorms, tornados, and tropical storms. At the other extreme, at times and places where it is dry, global warming will lead to increased drought intensity, higher temperatures, and more and stronger forest fires. Subtropical regions such as the American West, the Mediterranean region, Australia and parts of Africa are expected to be particularly hard hit by global warming. Because of earlier spring snowmelt and retreat of the glaciers, freshwater supplies will fail in many locations, as summers will be longer and hotter.

One difficulty in writing about climate change is that the situation is changing so rapidly — and is occurring much faster than even most scientists predicted just a short time ago. London-based journalist and author Gwynne Dyer wrote in February 2008 that this acceleration seems to be increasing at breakneck speed.

Dyer cited the latest issue of the journal *Science* that predicted climate change may cost southern Africa more than 30 percent of its main crop, corn, by 2030.[1] "No part of the developing world can lose one-third of its main food crop without descending into desperate poverty and violence," he wrote. In the developed world, Dyer notes, Australia is already in trouble because of its falling agricultural production. Still, any real action to address the problems continues to move at a snail's pace.[2]

"[T]here is no consensus on the best measures to deal with the problem, even among the experts, and the general public still does not grasp the urgency of the situation," according to Dyer.

"The two Democratic candidates for the U.S. presidency promise 80 percent cuts in emissions by 2050, and Republican Sen. John McCain promises 50 percent cuts by the same date, and nobody points out that such a leisurely approach,

1 *"Prioritizing Climate Change Adaption Needs for Food Security in 2030,"* Science, Feb. 1, 2008.
2 Gwynne Dyer, *"Panic in the Trenches,"* Feb. 1, 2008.

applied in every country, condemns the world to a global temperature regime at least 5.5 to 7 degrees Fahrenheit warmer than today," Dyer asserts, adding, "Few people are aware that these higher temperatures will prevent pollination in many major food crops in parts of the world that are already so hot that they are near the threshold, and that this, combined with shifting rainfall patterns, will cause catastrophic losses in food production."

Dyer interviewed a "couple of dozen" senior scientists, government officials, and other experts worldwide "and not one of them believes the forecasts on global warming issued by the Intergovernmental Panel on Climate Change" in 2007. "They think things are moving much faster than that." Dyer said the IPCC report "took no notice of recent indications that the warming has accelerated dramatically. While [the report] was being written, for example, we were still talking about the possibility of the Arctic Ocean being ice-free in late summer by 2042. Now it's 2013.

"Nor did the IPCC report attempt to incorporate any of the 'feedback' phenomena that are suspected of being responsible for speeding up the heating, like the release of methane from thawing permafrost. Worst of all, there is now a fear that the 'carbon sinks' are failing, and in particular that the oceans, which normally absorb half of the carbon dioxide that is produced each year, are losing their ability to do so."

While significant progress on combating the problem stalls, climate change is in the fast lane.

"[W]hile the high-level climate talks pursue their stately progress toward some ill-defined destination, down in the trenches there is an undercurrent of suppressed panic in the conversations," Dyer wrote. "The tipping points seem to be racing toward us a lot faster than people thought."

And the global downturn notwithstanding, the alarm bells keep ringing. A 2009 study sponsored by a consortium of U.S. government agencies reported that warming tied to higher CO_2 "is largely irreversible for 1,000 years after emissions stop."[1]

MSNBC reported that "[e]ven if the world can cap carbon dioxide emissions tied to global warming, expect to see droughts and sea level rise that span centuries, not just decades," according to the study.[2]

"People have imagined that if we stopped emitting carbon dioxide the climate would go back to normal in 100 years, 200 years; that's not true," lead author Susan Solomon said.

"Climate change is slow, but it is unstoppable," said Solomon, a researcher at the National Oceanic and Atmospheric Administration's Earth System Research

---

1 Jo Williams, *"Global Warming Study: U.S. Has Already Started Changing,"* Political Intelligence, Boston Globe, June 16, 2009. Report available at www.globalchange.gov/publications/reports/scientific-assessments/us-impacts.

2 *"Expect 1,000-Year Climate Impacts, Experts Say,"* MSNBC, Jan. 26, 2009.

Laboratory in Boulder, Colo. "All the more reason to act quickly, so the long-term situation doesn't get even worse."

Gerald Meehl, a senior scientist at the National Center for Atmospheric Research, said the research points out that "the real concern is that the longer we wait to do something, the higher the level of irreversible climate change to which we'll have to adapt."

The peer-reviewed study concludes that if CO_2 is allowed to peak at 450-600 parts per million [the air contained 280 ppm before the industrial revolution and is 385 ppm today], the results would include persistent decreases in dry-season rainfall that are comparable to the 1930s Dust Bowl in zones including the U.S. southwest, southern Europe, Africa, eastern South America and western Australia.

The study found that warmer climate also is causing expansion of the ocean and that factor alone is likely to lock in a 1.3 to 3.2 foot sea level rise by the year 3000 if CO_2 peaks at 600 ppm.

"We presented the minimum sea level rise that we can expect from well-understood physics, and we were surprised that it was so large," Solomon said.

She also noted that while global warming has been slowed by the oceans, which absorb carbon, that positive effect will wane over time and eventually oceans will actually warm the planet by giving off their accumulated heat to the air.

"[T]here would be changes in snow (to rain), snow pack and water resources...," according to Kevin Trenberth, head of climate analysis at the National Center for Atmospheric Research. "The policy relevance is clear: We need to act sooner ... because by the time the public and policymakers really realize the changes are here it is far too late to do anything about it. In fact, as the authors point out, it is already too late for some effects."

Chapter Four — Population

'Famine seems to be the last, the most dreadful resource of nature. The power of population is so superior to the power in the earth to produce subsistence for man, that premature death must in some shape or other visit the human race. The vices of mankind are active and able ministers of depopulation. They are the precursors in the great army of destruction; and often finish the dreadful work themselves. But should they fail in this war of extermination, sickly seasons, epidemics, pestilence, and plague, advance in terrific array, and sweep off their thousands and ten thousands. Should success be still incomplete, gigantic inevitable famine stalks in the rear and with one mighty blow levels the population with the food of the world.'

— British economist Thomas Robert Malthus (1766–1834)

A rapidly growing population will not be sustainable with a rapidly changing climate. According to J.R. McNeill in his essay "Historical Perspectives on Global Ecology":

Indeed, demographers now expect the human numbers will peak in half a century or so, at about 9 or 10 billion. The demographic pattern of the recent past is not only bizarre in light of the past, but will one day appear so in light of the future as well. Whether global population will stabilize at 9-10 billion, or enter into an era of decline or of unstable oscillations, is anyone's guess. We may safely say, however, that as with economic history, so with demographic history: the period since 1950 has been, in the perspective provided by longer sweeps of time, unimaginably bizarre.[1]

---

1 J.R. McNeill, essay *Historical Perspectives on Global Ecology*, Global Survival, editors Ervin Laszlo and Peter Seidel (New York: SelectBooks, Inc., 2006), p. 195.

Adds Richard D. Lamm in his essay, "Governance Barriers to Sustainability":

> There is a new public policy question that all nations must ask themselves: How many people can live satisfied lives in your nation? In the United States, our own birth rates will stabilize at only 340 million, but with immigration we will double, and then double again, the size of the nation. Demographic policy is a new and previously unimaginable issue that has thrust itself upon policy matters.
>
> The past century was unique. Perpetual growth is mathematically impossible in a finite space such as the Earth. Assume a starting figure and a growth rate — the rate is more important than the starting level — and I can show when the assumption of continued growth becomes absurd, by any standard you may choose. This is not a theoretical exercise. Current growth reaches the absurd very fast. As the United Nations Statistical Office pointed out in 1992, world population would reach 694 billion people in 2150 if the current fertility and mortality rates were maintained.
>
> Those limits to growth are usually ignored or denied, even by people who profess to be pursuing sustainability. The world has become addicted to growth, and those who profit from it are quick to deny that it must sometime stop.[1]

And according to Samuel Huntington in *The Clash of Civilizations and the Remaking of World Order*:

> Large populations need more resources, and hence people from societies with dense and/or rapidly growing populations tend to push outward, occupy territory, and exert pressure on other less demographically dynamic peoples. Islamic population growth is thus a major contributing factor to the conflicts along the borders of the Islamic world between Muslims and other peoples. Population pressure combined with economic stagnation promotes Muslim migration to Western and other non-Muslim societies, elevating immigration as an issue in those societies.[2]

There is an indisputable fact — fundamentalist Christians, the Muslim world, much of Africa, and ultra-orthodox Jews all have extremely high birth rates. While the one-child policy in China has reduced population growth, the sheer size of the country adds greatly to world population.

And authors Daniel Burstein and Arne de Keijzer write in *Big Dragon: China's Future*:

> Its ever-expanding population forces China to feed not just a quarter of the world's people, but an *additional* 13 million to 15 million mouths each year. In a single decade, China will add as many people as live in

---

1 Lamm, ibid., p. 222.
2 Samuel P. Huntington, *The Clash of Civilizations and the Remaking of World Order* (New York: Simon & Schuster Inc., 1996), p. 119.

Japan today. Within a generation, it will likely have added the equivalent of a whole United States.

China will need as much as 300 million tons of grain from world markets by 2030 — demand that would be far greater than all the global stocks now available for world export. A Chinese shortfall ... could empty the world's cupboards and even cause global famine.[1]

According to David Hale of Zurich Kemper Investments, "In 35 years, China will have a staggering 400 million people over the age of 65, compared to 100 million right now."[2]

It is little wonder that fears of resource scarcity, environmental degradation, and ethnic, religious, or racial conflicts due to rapid population growth often dominate discussions of national and international security, according to K. Bruce Newbold, director of the McMaster Institute of Environment and Health at McMaster University. In his book *Six Billion Plus: World Population in the Twenty-First Century*, he discusses what is at stake. Newbold observes that the world's population started to grow markedly beginning in the 1600s, as life expectancy slowly increased with improvements in commerce, food production, and nutrition. Even by 1800, the world's population was only one billion. By 1900, world population was approximately 1.7 billion, increasing to 2 billion by 1930. The mid-20th century saw unprecedented population growth, reaching 3 billion by 1960, growing to 4 billion by 1974. The fifth billion was reached just 12 years later, and by mid-2005, the total population was 6.48 billion.[3]

The number of cities in the developing world with populations in excess of one million will jump from 345 in 2000 to 480 by 2015. The number of megacities (cities with populations in excess of 10 million) has also grown from 8 in 1985 to 18 in 2000, and is projected to grow to 22 by 2015. Most of these new megacities will be in the developing world, and megacities will be home to an increasing proportion of the world's population. The growth of urban areas, driven by natural increase, net rural-to-urban migration, and urban reclassification, provides the raw ingredients for conflict.[4]

Newbold shows that in countries where fertility rates have dropped quickly, the young age structure of the population will ensure growth for the next two to three decades anyway. In other words, a huge proportion of the world's population has not avoided having children. Instead, these individuals *are* children. Consequently, a total world population of 7.9 billion by 2025 cannot be avoided, and most projections place world population between 7.3 and 10.7 billion by 2050, with nearly all of this growth occurring in the developing world.

---

1 Daniel Burnstein and Arne de Keijzer, *Big Dragon: China's Future: What It Means for Business, the Economy, and the Global Order* (New York: Simon & Schuster, 1998), p. 171.

2 Ibid., p. 278.

3 Newbold, p. 6.

4 Ibid., p. 6.

Pronounced differences in population growth and structure separate the developed and developing worlds; the geographic distribution is becoming more unbalanced. Rapid population growth in the second half of the 20th century has meant that the share of the world's population residing in the developing world climbed from 68 to 81 percent. Regionally, North America and Europe represent only 16 percent of the current population. According to U.N. projections, the percentage residing in the developing world will grow to 86 percent by 2050.[1]

In Europe, governments may choose to increase immigration as the native population ages, but it will do so with great risk. Europe has not, in the past, been a major destination for immigrants (excepting for short-term work programs), and current immigration numbers are insufficient to reverse population decline, while further increases in immigration levels may result in ethnic confrontation. Most European countries have imposed strict immigration policies, and some have actively encouraged their foreign-born populations to leave.

The U.S. government has taken a different approach, on the one hand removing some of the incentive to flee impoverished and overpopulated lands by directing aid money through its own Agency for International Development (USAID), reflecting its own concerns and policy goals. Newbold says that, largely driven by concerns that rapid population growth in the developing world threatened U.S. security through trade, political conflict, immigration, or damage to the environment, USAID has been the largest single donor to family-planning programs.[2] U.S. funding for overseas family planning (or population control) programs is a divisive issue between Republicans and Democrats.

The spread of disease is an increasing concern. Rapid population growth and urbanization challenge governments to provide public infrastructure including clean drinking water, sanitation, appropriate housing, public education and other programs, let alone basic health care services. Mass transit and long distance air travel convey diseases as well as their hosts, potentially spreading germs across the continent and around the world in a matter of hours.[3] There are also an increasing number of cases of societal rejections of immunization, based on religious grounds or on a lack of trust in the government.

As an economic process, international migration is motivated by a combination of "push" factors in the place of origin, including poor employment prospects, large populations, and low wages. Most of those who leave home are emigrating from Asia, North Africa, and Latin America, and heading for the United States, Canada, Australia, Western Europe, Scandinavia, and Russia, where even a low wage is better than they could find at home.

---

1 Ibid., p. 9.
2 Ibid., p. 37.
3 Ibid., p. 57.

Increasing immigration increases the threat of ethnic, racial, and social insta-bility while creating a pool of workers who will accept extremely low wages and who compete for jobs with the native born.[1] This competition from foreigners can stoke stronger feelings of national identity and fuel concerns over cultural and linguistic protection, as well as differences between foreigners and the na-tive born.

Both Europe and the United States face enormous political pressures when it comes to controlling immigration. Employers desire cheap labor, while native-born workers see their power and livelihood threatened. Globalization of the economy opens the door to greater trade and capital flows and increases demand for cheap labor within industrialized countries.

The Arab world has migration problems of its own. With the exception of Jordan, where more than half of the population is Palestinian, the admission and integration of Palestinians by other countries has been poor, largely due to host governments' fear that removing the refugee label would destroy chances for recreating a Palestinian state. But the presence of refugees may inflame tensions between countries or ethnic groups. In Lebanon, the delicate balance between Muslims and Christians has prevented Palestinian refugees from naturalizing for fear of upsetting the political balance. In other cases, fighters or militia members often use refugee camps as bases, promoting instability both within and outside the camps.[2]

Science has investigated the complexity and interconnectivity of ecologi-cal systems and its implications for the population. In the past, the Earth's en-vironmental systems were regarded as stable and resilient to human tampering. Instead, there is mounting evidence drawn from observation of ocean currents, ozone depletion, and fish stocks that environmental systems are not stable in the face of human action.

It is questionable whether some countries, such as China, Egypt, and India, have the resources and economic ability to sustain their populations indefinitely even if population growth were to cease immediately; and Uganda's population is projected to increase by 310 percent between 2007 and 2050. Writing for the Worldwatch Institute in 1995, Lester Brown forecasted that a combination of rising standards of living, increased consumption, loss of cropland, and declining water resources, among other factors, would force China to turn to world mar-kets to purchase foodstuffs. The problem lies in the country's expected demand for grains, which Brown projected would exceed total world output, driving up prices globally and weakening the ability of smaller, poorer countries to purchase their requirements.[3]

---

1 Ibid., p. 135.
2 Ibid., p. 145.
3 Ibid., p. 176.

Despite the so-called "green" revolution, it is unrealistic to expect future increases in crop yields and agricultural output similar to those observed during its peak. In fact, declining rates of growth in agricultural output and increasingly marginal returns on the application of additional inputs, such as fertilizers, have recently been observed: the nascent "green revolution" is already ending.[1]

In his 1994 article "The Coming Anarchy," journalist Robert Kaplan paints a dire picture of the world's future in which peripheral states, robbed of their economic power by globalization, poor leadership, and environmental decay, disintegrate into smaller units defined by ethnicity or culture and ruled by warlords and private armies. In this scenario, violence and conflict would become the norm.[2]

Ecological as well as economic marginalization may prompt migration or population displacement, creating what some authors have called "environmental refugees." Egypt provides a useful example of the links between these processes. Confined to less than 3 percent of the country's territory, Egypt's population of 74 million is strung along the Nile River. Heavily dependent on its waters, Egypt has been taxed economically by population growth (the total fertility rate remains greater than 3, and the country's population will double in approximately 35 years under current conditions), contributing to low-level but ongoing conflict between Muslim fundamentalists and the more secular government. Egypt's weak economic position, along with local scarcities of land and water, has reduced its ability to pursue economic and agricultural development, or to provide institutions and services like health care and education for its population, reducing its ability to inspire the loyalty of citizens. As a result, the fundamentalist Muslim Brotherhood has stepped in, a group that has openly and violently challenged the authority of the Egyptian government. Opposing Egypt's secular government, it has cultivated grassroots support by operating social institutions, such as schools, clinics, hospitals, and charities, providing otherwise unavailable services to the poor. Simultaneously, fundamentalist groups have sought to weaken Egypt's government by attacking tourist operations, a key income-generating industry, such as the July 2005 bombings of Red Sea resorts. Finally, differences in religious viewpoints have also promoted social segmentation as some of the more radical fundamentalist groups have isolated themselves from mainstream social and religious institutions. The development of fundamentalist views has placed increasing pressure on Coptic Christians, who, representing approximately 10 percent of the population, face discrimination.[3]

With projections that known world oil supplies will likely peak within the next 20 years (some experts argue that oil has peaked already), oil is likely to

---

1 Ibid., p. 184.
2 Robert D. Kaplan, *"The Coming Anarchy,"* The Atlantic Monthly, February 1994.
3 Newbold, ibid., p. 196.

remain a "prize" resource much fought over in the coming years. This form of re-source capture — the decreasing quality or quantity of a resource interacts with population growth and increasing consumption, encouraging groups to control that resource through trade or military conquest — can also be extended to re-newable resources, such as cropland, forests, or fresh water.[1]

The growth of urban populations worldwide, especially large, dense, and poverty-stricken populations, means that the potential for conflict arising from resource scarcity is also escalating.[2]

Urban violence, whether it is directed toward the state or takes place be-tween groups such as ethnic or religious violence based on perceived differences between groups, is likely to become increasingly common. The current relative quiet of urban populations in the developing world may not last, given increas-ing poverty and the increased strength and reach of organized crime in cities throughout the developing world. States frequently respond to crises and unrest with increased repression, which only amplifies the likelihood of conflict. The poor, underemployed, or unemployed young provide ready fodder for uprisings, at the same time that the balance of power shifts from the government to other groups, including crime gangs and religious fundamentalist movements. In this way, urban violence is not just a developing-world issue. Following the bomb-ings of London's subway in July 2005, concerns were voiced not just over unem-ployment and restlessness among first-generation immigrant youth and their po-tential for violence against the state, but also with second- and third-generation immigrants who have been economically constrained within British society by class and race barriers. The U.S. government was also concerned with this group: as British citizens, they would not require a visa when traveling to the United States.

Although globalization has the potential to unite groups, it also comes with the potential to further separate the haves and have-nots as groups or countries are left behind. Already, globalization has created winners and losers within the existing world order. Portions of the developing world have been left behind, vic-tims of poor resources, poor government, or investments that have simply passed them by. The gap between developed and developing countries, whether mea-sured by investment, capital, level of education, income, or productivity, is wid-ening and is expected to continue to grow.[3]

Newbold continues, "Due to globalization, which has tended to bypass many of the poorest countries, as well as the increased power of warlords, crime gangs, drug cartels, and guerrilla groups, future conflict may be 'borderless,' failing to conform to existing notions of interstate or intrastate conflict, with influence

---

1 Ibid., p. 198.
2 Ibid., p. 202.
3 Ibid., p. 207.

exerted not by a state (if states continue to exist) but by ethnic groups or clans. At the extreme, territory will not be marked by traditional state borders but instead will be defined by shared ethnicity or some similar construct, such as clan or tribe, resulting in a shifting map as power or allegiances change."

The developed world will not be immune to the consequences of environmental scarcity. In the southwestern United States, for example, states increasingly compete for scarce water resources. In Texas, the state population is nearly 21 million, a number expected to double within 50 years. El Paso is already experiencing water shortages, and water in the Rio Grande, which is a primary source of water in the area, is nearly all being used, with only a trickle reaching the ocean. In other areas, the Ogallala Aquifer is overpumped and stressed. In Texas, surface water resources are public but underground water is private; privatization of underground water rights and "water farms" allow water to be pumped without regard for neighbors, a form of resource capture. Indeed, Texas law recognizes it as the "rule of capture." Widespread public concerns have been generated that water pumped for profit and sold to the highest bidder, usually cities or industry, which could threaten supply in some areas, marginalizing those who cannot afford it and potentially disrupting or destroying irrigated agriculture in the Southwest, creating a situation reminiscent of the great Dust Bowl migration portrayed in John Steinbeck's *The Grapes of Wrath*. On a continental scale, the need to quench the thirst of America's southwestern states has several entrepreneurs and Canadian provinces, including Newfoundland, exploring the idea of bulk water shipments. Home to some 9 percent of the world's freshwater supplies, Canada is unsure where it stands on the idea of exporting bulk water, although it has so far refused to sell bulk water shipments, citing safeguards within the North American Free Trade Agreement (NAFTA). Under NAFTA, water is exempted from trade rules that require two-way trade in commodities. If Canada were to start exporting bulk water, however, free trade provisions would be invoked, requiring the trade to continue and restricting the government's ability to limit its flow, allowing provincial and state governments to challenge the federal position.[1]

Globalization of the world's economy may be touted as the harbinger of a new economic order that can lift nations out of poverty.[2]

"Yet, it is more likely that large regions of the world will be left out of the new international arrangement," Newbold asserts. "Geography matters. Africa, the Indian subcontinent, and parts of Asia will become marginalized, characterized by a lack of international investment, poor leadership, few resources, and distance from markets. Unless the developed world is prepared to extend aid and investment, reversing the declines in international aid noted over the past

---

1 Ibid., p. 209.
2 Ibid., p. 223.

few years, these regions are likely to be pulled into a reinforcing cycle of violence and poverty. Conflict will not take the traditional form of state versus state for control of territory. Instead, conflict will be subnational and defined by local re-source scarcities, with group-identity conflicts focusing on ideological or ethnic differences — a "we" versus "them" mentality — or local insurgencies and civil strife. National governments will cease to operate, replaced by warlords and pri-vate armies controlling shifting territories."

China has proclaimed that it will continue its one-child policy, which limits couples to having one child, through 2010. The policy was established by Chinese leader Deng Xiaoping in 1979 to limit China's population growth. Chinese offi-cials stated recently that the policy has prevented 400 million births — consider-ably more births than the entire population of the United States which was 302.2 million people in mid-2007. The birth rate in China was 6.2 in the early 1950s and had declined to 1.8 in 2007. Even so, China still adds millions of people to the population each year. Although the drop has been remarkable, many demog-raphers and varied scientists still question whether China can feed this growing population during a time of global warming, increased energy use and consump-tion, and environmental degradation.

The one-child policy has brought howls of disapproval from primarily West-ern Christian pro-life groups and Western human rights organizations.

Critic Wendy McElroy, appearing on *Fox News Views* in 2002, said the policy was "arguably the greatest bioethical atrocity on the globe."

The anti-abortion groups have lobbied hard against the Chinese policy. One such group is the Society for the Protection of Unborn Children based in the United Kingdom. According to the group, most Western governments support the policy. "Most of the opposition to the policy has come from pro-life organiza-tions like (the) SPUC," according to the group. "SPUC has in recent years taken a leading role on the issue, working closely with expert groups, in particular the Population Research Institute and the Catholic Family & Human Rights Insti-tute. Pro-lifers are therefore urged to help support the work of this international pro-life coalition against the one-child policy by lobbying politicians, govern-ments, and human rights groups."

The official-sounding Population Research Institute is a self-described pro-life educational organization "dedicated to protecting and defending human life, ending human rights abuses committed in the name of family planning, and dis-pelling the myth of overpopulation." The PRI has in past years received funding from The Lynde and Harry Bradley Foundation, Inc., which has been described as "the country's largest and most influential right-wing foundation." The presi-dent is Catholic convert Steve Mosher.[1]

---

1 Steven W. Mosher, *"China's One-Child Policy,"* Population Research Institute, September-October 2001.

While global warming has its plethora of deniers, the PRI denies overpopulation in the face of overwhelming evidence — and the U.S. government, especially under Reagan and George W. Bush, responded positively. Any reasonable demographer understands that China — even with the one-child policy — is one harvest away from mass starvation. Once again, many consider concern about overpopulation as a plot by the U.N. Mary McCarthy, writing in the *Catholic Herald* in 2003, stated: "Each year in China, millions of babies are killed by force." She adds that the United Nations Population Fund "claims that one of its priorities is to protect young people, but as Population Research Institute representatives found when they visited China in 2001, UNFPA officials are sharing offices with the very people who kill an unimaginable number of children each year in the name of 'family planning.'"

Another target of vitriol is the U.N.'s family planning efforts in Afghanistan, which has a whopping world-leading fertility rate of 7. "These same women said that they want clean water, immunizations for their children, and the chance to learn more about natural family planning," McCarthy continues. "The majority of them want more children."[1]

These groups have been received warmly — beginning with the Reagan Administration and forcefully followed up during the George W. Bush Administration. In 2004, the U.S. State Department stated that "China is using coercion to enforce its 'one-child' per couple policy."

Arthur Dewey, the former U.S. assistant Secretary of State for the bureau of population, refugees and migration, reiterated the Bush Administration's stand which had "strongly and absolutely" opposed the "practice of coercive abortions and sterilizations wherever they occur." The answer? Pull funding from the U.N. Population Fund.[2]

Newbold, who actually studies population growth and has written widely on the subject, responded: "While Chinese fertility rates would have likely declined anyways (in tandem with its rapid economic development), had it not instituted the policy when it did, world population growth would have been much greater, and China's population would have been much larger than it currently is." He added that the policy has shifted China's demographic structure "dramatically" in comparison with other developing nations.

The Population Reference Bureau, a respected organization dealing with population, reported in 2007 that the global population will increase by 40 percent by 2050, with 49 percent of the growth coming from less developed countries. Excluding China, the rate would be 61 percent. China has taken the lead in curbing unsustainable population growth — and, in the process, had incurred

---

1 Mary McCarthy, *"China's One-Child Policy,"* Catholic Herald, March 20, 2003.
2 *"China's 'One-Child' Policy Coercive, State's Dewey Says,"* U.S. Department of State, Dec. 14, 2004.

the wrath of the former Bush Administration and the religious groups who supported him.

One fact is beyond dispute — family size decreases with the level of education and access to family planning programs. The United States, which was already ducking any serious response to climate change under the Bush Administration, also denied its evil twin — overpopulation. The country and the world do so at its own peril.

The U.S. will not be immune from population increase even though the fertility rates today are at replacement levels of 2.1 and has put an interesting spin on the current immigration debate. The increase of immigrants from Mexico will substantially increase the U.S. population because of cultural and religious factors. Mexicans have larger families — with a fertility rate exceeding 4. The hard fact is that low-wage immigrants will also have the largest families to support which will put more stress on government assistance programs. The current population of 302.2 million is expected to increase to 349.4 million by 2025.

According to the Population Reference Bureau's annual report for 2007, the world entered the 20th century with a population of 1.6 billion people. We entered the 21st century with 6.1 billion people. And in 2007, world population was 6.6 billion. This increase in the size of the human population is unprecedented.[1]

And nearly all of the growth is occurring in the less developed countries. Currently, 80 million people are being added every year in less developed countries, compared with about 1.6 million in more developed countries. It is remarkable that, despite many new developments over the past 50 years, one fact looks very much the same: Populations are growing most rapidly where such growth can be afforded least. The Middle East and North Africa regions have experienced the largest increase in life expectancy since the late 1950s: from 43 years to 70 years and continue to have exceedingly high birth rates.

The United States is the only developed country experiencing significant population growth for two main reasons: The U.S. has a higher fertility rate than other developed countries, and the U.S. has more net migration than other developed countries. The United States receives about 20 percent of the world's international migrants, but the U.S. accounts for just 5 percent of the world's population. Foreign-born couples tend to have more children than U.S.-born couples. Foreign-born residents are in their prime childbearing years, and immigrants often come from countries where larger families are more common.

For the non-Hispanic white population, the ratio of births to deaths is nearly equal: almost 1:1. Demographers would say this represents no natural increase in population size. Hispanics, on the other hand, have eight births for every death, leading to a significantly larger growth rate than that for non-Hispanic whites.

---

1 *"World Population Data Sheet, Annual Report,"* Population Reference Bureau, 2007.

The United States is the largest contributor of total carbon dioxide emissions and has one of the highest per capita rates. The U.S. per capita emission rate has risen from 19.2 metric tons per person to 19.9 metric tons between 1990 and 2002, according to the World Resources Institute.[1]

Per capita use also has gone up in China, rising from 2.2 to 2.9 metric tons between 1990 and 2002. China was expected to surpass the U.S. in total carbon dioxide emissions by 2009.

The population of more developed countries is 1.2 billion; less developed, 5.4 billion. In 2005, about 191 million people — 3 percent of the world's population — were international migrants, according to U.N. estimates.

Between 1995 and 2000, around 2.6 million migrants per year moved from less developed to more developed regions. More than one-half of these settled in the United States and Canada. About 40 percent of international migrants moved from one less developed country to another.

Immigration has a major effect on the size, distribution, and composition of the U.S. population and its role has increased because national birth and death rates are relatively low. Immigration contributed at least a third of the total population increase between 1990 and 2000, as the number of foreign-born U.S. residents rose from almost 20 million to over 31 million. The number of foreign-born persons (the first generation) is projected to rise, from 31 million in 2000 to 48 million in 2025, and the foreign-born share of the U.S. population is projected to increase from 11 to 14 percent. Accordingly, the number of second-generation Americans — the children of immigrants — will continue to increase.

In 2000, first- and second-generation Americans were about 21 percent of the U.S. population. If net legal and illegal immigration averages just 820,000 per year, first- and second-generation Americans are projected to be about one-third of the U.S. population in 2025.

In Africa, 41 percent of the population is under age 15; in Europe, the number is 16 percent.

The world is on the verge of a shift: from predominantly rural to mainly urban. In 2008, more than half of the world's people lived in urban areas. By 2030, urban dwellers will make up roughly 60 percent of the world's population. In North America, Europe, and Latin America and the Caribbean, more than 70 percent of the population is already urban; but in Africa and Asia, less than 40 percent of this population is urban.

Contrary to popular belief, the bulk of urban population growth is likely to occur in smaller cities and towns of less than 500,000. Urban people change their environment through their consumption of food, energy, water, and land. In turn, the polluted urban environment affects the health and quality of life of the urban population.

---

1 *"Annual Carbon Dioxide Inventory Report,"* World Resources Institute, 2004-2005.

People who live in urban areas have very different consumption patterns than residents of rural areas. For example, urban populations consume much more food, energy, and durable goods than rural populations.

By extension, the energy consumption for electricity, transportation, cooking, and heating are much higher in urban areas than in rural villages. For example, urban populations have many more cars per capita than rural populations.

Urban consumption of energy also creates "heat islands" that can change local weather patterns and weather downwind from these islands. The heat island phenomenon is created as cities radiate less heat back into the atmosphere than rural areas, making cities warmer than rural areas. These heat islands also trap atmospheric pollutants. Cloudiness and fog occurs more often. Precipitation is 5 to 10 percent higher in cities, and thunderstorms and hailstorms are much more frequent. Urbanization also affects environments beyond the city. Regions downwind from large industrial complexes see increases in the amount of precipitation, air pollution, and the number of days with precipitation. Urban areas also affect water runoff patterns. Not only do urban areas generate more rain, they reduce the infiltration of water and lower the water tables. This means that runoff occurs more quickly with greater peak flows. Flood volumes increase, as do floods and water pollution downstream.

But global urban expansion takes less land than land lost every year to agriculture, forestry, and grazing, or to erosion and salinization.

Population growth, as should be obvious by now, is pushing the finite world closer to unsustainability.

CHAPTER FIVE — BUSINESS AS USUAL

The fundamental business of the country is on a sound and prosperous basis, said President Hoover. No buildings were burned down, no industries have died, no mines, no railroads have vanished, crooned Arthur Brisbane. The great comforters. There, there my children. Try and catch a little sleep. Mother is near.
— "The Market Goes South," *The New Yorker*, 1929

They walked through the streets wrapped in the filthy blankets. He held the pistol at his waist and held the boy by the hand. At the farther edge of the town they came upon a solitary house in a field and they crossed and entered and walked through the rooms. They came upon themselves in a mirror and he almost raised the pistol. It's us, Papa, the boy whispered. It's us.
— *The Road*, Cormac McCarthy

While television ads and newspapers are increasingly filled with industry's new-found conversion to "green" technology and practices, analysts and climate scientists are skeptical that any substantial changes are likely to occur. Global warming and American-style free market capitalism, dependent as it is on short-term constant growth with no feasible alternatives to fossil fuels waiting in the wings, are heading toward an inevitable collision. Neither is likely to apply the brakes or switch tracks before the inevitable crash. The Earth is finite; capitalism is not.

Cut fossil fuel use? It's not possible in any meaningful way for a number of reasons. Bend the Earth to industry's will? In combat involving man versus nature, always bet on nature in the long term.

In 1992, Francis Fukuyama, a political scientist at Johns Hopkins University, published *The End of History and the Last Man*, arguing that historical progression had led towards secular free-market democracy. This book was interpreted by many free-marketeers as the inevitability of unlimited free markets and globalization. With the fall of the Berlin wall, capitalism was heralded as a God-ordained global norm. However, Professor Fukuyama woefully underestimated two serious problems looming on the horizon: radical climate change and cultural considerations.[1]

In defense of Professor Fukuyama, however, he said future economic systems may more closely resemble the social democracies in the European Union rather than U.S.-style capitalism.

Environmentalists have since pointed out that the relentless growth required by capitalist economies will conflict directly with the Earth's already scarce resources. The impending environmental meltdown will require a radical alteration of the capitalist system as it struggles to deal with these changes. However catastrophic these changes may negatively alter life on the planet, corporate interests with almost incomprehensible amounts of money at stake have not and will not go silently into this dark night. Business interests have been remarkably successful at molding public opinion — even though it is against the best long-term interests of the average American — which they have honed successfully over the past century. While the scientific community has provided ample information on the serious consequences of rapid climate change, the business community continues to sow the seeds of doubt about climate change — and has done so for decades. Scientists have and will continue to be disemboweled in the meat grinder of national politics ... until the window for taking effective action has long passed. Business as usual will continue — under Democrats or Republicans — because politicians are also motivated by short-term returns (election and re-election) and corporate financial backing.

So what will business as usual look like in the year 2030? A good guide is provided by the U.S. Department of Energy's International Energy Outlook for 2007 published by the Energy Information Administration.[2] The DOE is quick to add the disclaimer that "information contained herein should be attributed to the Energy Information Administration and should not be construed as advocating or reflecting any policy position of the Department of Energy or of any other organization." It should also be noted that this projection was made before the global economy went into a tailspin. Definitive figures for 2008 have not yet been published.

The report graphically demonstrates increases in energy use by developing nations — especially China and India. When energy *increases* are considered, the developing nations seem to be embarking on an unsustainable increase in carbon dioxide emissions. But when *total energy use* is considered, the United States is

---

1   Francis Fukuyama, *The End of History and the Last Man* (New York: Avon Books, Inc., 1992).

2   *"U.S. Department of Energy's International Energy Outlook,"* Energy Information Administration, 2007.

still, by far, the biggest pig at the trough. The world's three largest economies, locked in global competition, will not cede ground to one another. China and India argue — with some moral and ethical justification — that they should not be denied the economic benefits of expansion enjoyed so long in the West. The United States is loath to give up its position as the world's economic leader. This heated competition in a finite and already overstressed environment is a recipe for disaster that will take only decades, if that, to manifest itself. The public relations firms will continue to babble and fingers will continue to be pointed to the detriment of both developed and developing countries alike.

Publicly, scientists continue to put on a brave face — if immediate, serious action is taken, substantial progress can be made before the engine that controls the global climate overheats and causes unimaginable disaster. Privately, these scientists are less optimistic, even fatalistic — they are well aware that the window is slamming shut. However, as most scientists and environmentalists agree, publicly expressing pessimism serves no good purpose. Optimism leads to action. Despair leads to lethargy and a sense of doom. These attitudes are also the antithesis of the conventional American "can do" Babbittry central to the American character. The facts are less appealing. Changes are occurring faster than computer models predicted. Technology, often cited as an elixir for these predicaments, has its limits and no serious replacements for fossil fuels are in the pipeline despite the best efforts of both government and the corporate world to assuage a nervous public. No infrastructure to replace fossil fuels can possibly be put in place before the planet reaches critical tipping points. The U.S. will balloon its national debt by bailing out the financial sector with an influx of an estimated $3 trillion. This unexpected financial burden will inhibit both industry and government from spending the necessary funds to prepare for the climate catastrophes that loom on the horizon.

For all of the lip service given to the welfare of children and grandchildren — the bicycle helmets, the Amber alerts, and stranger danger — the Woodstock Nation is on its way to becoming the most vilified of all generations, driven by unprecedented greed and consumption, debt, and accelerating environmental degradation. It is difficult to ignore the hypocrisy of a generation which once showed so much promise and inaugurated Earth Day. Unfortunately, this culture of greed, conspicuous consumption, and narcissism has been passed down to a significant number of our children and grandchildren. It's not about "us" — it's about "me."

The following projections by the DOE, published yearly, are not produced out of intellectual curiosity about energy usage worldwide. The report, which forecasts global energy use up to the year 2030, was mandated by the Department of Energy Organization Act of 1977 under the Carter Administration. The report purports to be "objective" and "policy-neutral."

The noteworthy projections are as follows:

• World marketed energy consumption was projected to increase by 57 percent from 2004 to 2030. Total energy demand in the non-OECD (Organization for Economic Cooperation and Development) countries increased by 95 percent, compared with an increase of 24 percent in the OECD countries. Total world energy use was projected to rise from 447 quadrillion British thermal units (BTU) in 2004 to 559 quadrillion BTU in 2015 and then to 702 quadrillion BTU in 2030. Global energy demand grows despite the relatively high world oil and natural gas prices that are projected to persist into the mid-term outlook.

• In all the non-OECD regions combined, economic activity — as measured by GDP in purchasing power parity terms — increased by 5.3 percent per year on average, as compared with an average of 2.5 percent per year for the OECD economies. [These figures are now obviously inaccurate because of the current global economic recession.]

• For the industrial sector, energy-intensive industries continue to expand more rapidly in the non-OECD countries, where *investors are attracted by lower costs and fewer environmental constraints* [italics added], than in OECD countries.

• For the non-OECD region as a whole, strong growth in demand for energy is projected in the buildings sectors, averaging 2.4 percent per year in the residential sector and 3.7 per year in the commercial sector.

• With robust economic growth projected for the developing non-OECD nations, transportation sector energy use increases by an average of 2.9 percent a year from 2004 to 2030, requiring extensive investment in the construction of transportation infrastructure (highways, fueling stations, airport facilities, rail systems, etc.) to support the fast-paced growth in demand spurred by passenger car travel.

• Fossil fuels (petroleum and other liquid fuels, natural gas, and coal) are expected to continue supplying much of the energy used worldwide. Liquids (petroleum) remain the dominant energy source, given their importance in the transportation and industrial end-use sectors.

• World use of petroleum grows from 83 million barrels equivalent per day in 2004 to 118 million barrels per day in 2030. Liquids remain the most important fuels for transportation, because there are few alternatives that can compete widely with petroleum-based liquid fuels. On a global basis, the transportation sector accounts for 68 percent of the total projected increase in liquids use from 2004 to 2030, followed by the industrial sector, which accounts for another 27 percent of the increase.

• To meet the increment in world liquids demand, total supply in 2030 is projected to be 35 million barrels per day higher than the 2004 level of 83 million barrels per day. OPEC contributes about 21 million barrels per day to the total increase and conventional liquids production in non-OPEC countries adds another 6 million barrels per day. Unconventional resources are projected to increase to 10.5 million barrels per day and account for 9 percent of total world liquids supply in 2030.

• Natural gas consumption increases on average by 1.9 percent per year, from 99.6 trillion cubic feet in 2004 to 163.2 trillion cubic feet in 2030. Rising world oil prices after 2015 increases the demand for — and then the price of — natural gas, as it is used to displace the use of liquids in the industrial and electric power sectors.

• Higher natural gas prices, in turn, make coal more cost-competitive, especially in the electric power sector.

• Coal is the fastest-growing energy source worldwide. World coal consumption is projected to increase from 114.5 quadrillion BTU in 2004 to 199.1 quadrillion BTU in 2030, at an average annual rate of 2.2 percent. Coal use jumped 17 percent between 2003 and 2004 in Asia, mainly due to China and India. Coal's share of total world energy use is projected to increase from 26 percent in 2004 to 28 percent in 2030.

• The electric power sector accounts for about two-thirds of the world's coal consumption throughout the projection period, and the industrial sector accounts for most of the remainder. China's industrial sector is projected to account for about 78 percent of the total net increase in industrial coal use worldwide. China has abundant coal resources, limited reserves of oil and natural gas, and a leading position in world steel production.

• World net electricity generation grows by 85 percent from 16,424 billion kilowatt-hours in 2004 to 30,364 billion kilowatt-hours in 2030. Coal and natural gas remain the most important fuels for electricity generation throughout this period.

• The renewables share of total world energy consumption is expected to rise from 7 percent in 2004 to 8 percent in 2030. Most is generated by hydroelectric facilities. Outside of Canada and Turkey, hydropower capacity is not expected to grow substantially in OECD nations because most hydroelectric resources in the region have already been developed or lie far from population centers.

• World carbon dioxide emissions continue to increase steadily from 26.9 billion metric tons in 2004 to 42.9 billion metric tons in 2030, an increase of 59 percent over the projection period.

• The year 2004 marked the first time in history that energy-related carbon dioxide emissions from the non-OECD countries exceeded those from the OECD countries — although by only about 8 million metric tons. Further, because the projected average annual increase in emissions from 2004 to 2030 in the non-OECD countries is more than three times the increase projected for the OECD countries, carbon dioxide emissions from the non-OECD countries in 2030, at 26.2 billion metric tons, are projected to exceed those from the OECD countries by 57 percent.

● Generally, countries outside the OECD have higher projected economic growth rates and more rapid population growth than OECD nations. The OECD countries have more mature national economies and population growth is expected to be slower. Energy use in the non-OECD region is projected to surpass that in the OECD region by 2010, and to be 35 percent greater than the non-OECD total in 2030.

● In the United States, dependence on relatively expensive domestic supplies of unconventional natural gas and imports of liquefied natural gas (LNG) is expected to increase over the projection period, and projected prices in the U.S. market thus tend to be at the high end of the range. In Russia and the Middle East, where domestic resources of conventional natural gas are both abundant and readily accessible, natural gas prices are among the lowest in the world.

● Coal's share of total world energy use climbed from 25 percent in 2003 to 26 percent in 2004. With oil and natural gas prices expected to continue rising, coal is an attractive fuel for nations with access to ample coal resources. Its usage is expected to increase to 28 percent in 2030. In particular, the United States, China, and India are well-positioned to displace more expensive fuels with coal, and together the three nations account for 86 percent of the expected increase from 2004 to 2030. Decreases in coal consumption are projected only for OECD Europe and Japan.

● Robust economic growth in many of the non-OECD countries is expected to boost demand for electricity to run newly-purchased home appliances for air conditioning, cooking, space and water heating, and refrigeration and to support the expansion of commercial services, including hospitals, office buildings, and shopping malls.

● The relative environmental benefits and efficiency of natural gas make it an attractive fuel choice for generation in many nations; however, higher oil and natural gas prices make coal the economic choice in the United States and non-OECD Asia, where coal resources are ample.

● Much of the growth in renewable energy consumption is projected to come from mid- to large-scale hydroelectric facilities in non-OECD Asia, and Central and South America.

● Over the 2004 to 2030 period, world real GDP growth is projected to average 4.1 percent annually primarily based on the growth prospects of China and India.

● Improved macroeconomic policies, trade liberalization, more flexible exchange rate regimes, and lower fiscal deficits have lowered [developing nations'] inflation rates, reduced uncertainty, and improved their overall investment climates. More microeconomic structural reforms, such as privatization and regulatory reform, have also played key roles. In general, such reforms have resulted in growth rates that are above historical trends in many of the emerging economies over the past 5 to 10 years. [This cheery outlook also tanked with the global economy.]

• A downturn in the U.S. housing sector has been the major source of weakening over the past year, and reductions in manufacturing output indicate that the slowdown has spread throughout the economy. At the same time, however, corporate finances have been healthy, and real nonresidential investment has remained robust. [This, of course, has also been proven false by the recent economic meltdown.]

• In reference case projections, the U.S. economy stabilizes at its long-term growth path by 2010. GDP is projected to grow by an average of 2.9 percent per year from 2004 to 2030 — slower that the 3.1-percent annual average over the 1980 to 2004 period — because of the retirement of the baby boom generation and the resultant slowing of labor force growth. [This has also proven to be false.]

• In Mexico, real GDP is projected to grow by an average of 3.6 percent per year from 2004 to 2030. Remittances from Mexicans working abroad (the U.S.) continue to grow rapidly, boosting domestic consumption. Mexico's industrial production follows, and is heavily influenced by, U.S. GDP growth and outsourcing of employment. Mexico's future growth is also more dependent on U.S. growth.

• According to the International Monetary Fund, structural impediments to economic growth still remain in many countries of OECD Europe, related to the region's labor markets, product markets, and costly social welfare systems. [Damn those welfare systems!] Reforms to improve the competitiveness of European labor and product markets could yield significant dividends in terms of increases in regional output.

• China, non-OECD Asia's largest economy, is expected to continue playing a major role on both the supply and demand sides of the global economy. The report projects an average annual growth rate of approximately 6.5 percent for China's economy over the 2004 to 2030 period. The country's economic growth is expected to be the highest in the world. The report is critical of inefficient state-owned companies and a banking system that is carrying a significant amount of nonperforming loans.

• India's economic growth is viewed as positive because it continues to privatize state enterprises and increasingly adopts free market policies. Average annual GDP growth in India over the 2004 to 2030 projection period is 5.7 percent. The report states, as in China: "Accelerating structural reforms — including ending regulatory impediments to the consolidation of labor-intensive industries, labor market and bankruptcy reforms, and agricultural and trade liberalization — remains essential to stimulate potential growth and reduce poverty in the medium to long term. With its vast and relatively cheap labor force, India is well positioned to reap the benefits of globalization."

• The rest of non-OECD Asia is robust in part because of increasing exports to China.

• Medium-term prospects for the Middle East remain favorable, given that a significant portion of the recent increase in oil revenues is expected to be permanent. [Oil prices have tanked recently but now seem to be on an upward path again.]

• The report puts world oil prices at $59 per barrel in 2030.

• Over the next 25 years, demand for petroleum and other liquid fuels is expected to increase more rapidly in the transportation sector than in any of the other end-use sectors. In the OECD countries, which are projected to remain the greatest users of energy for transportation, the transportation sector's share of total liquids demand is projected to rise from 58 percent in 2004 to nearly 63 percent in 2030. A primary factor contributing to the expected increase in energy demand for transportation is steadily increasing demand for personal travel in both the developing and mature economies. Increases in urbanization and in personal incomes have contributed to increases in air travel as well as increased motorization (i.e., more vehicles) in growing economies. As trade among countries increases, the volumes of freight transported by air and marine vessels is expected to increase rapidly over the projection period.

• Alternative fuels remain fairly expensive. Barring any widespread increase in penetration of new technologies, whether driven by policy changes or other factors, the world's use of alternative fuels in the transportation sector is expected to remain relatively modest through 2030 in both OECD and non-OECD countries.

• In the United States, the transportation sector continues to account for almost one-fourth of the country's total energy consumption. The United States is the largest user of transportation energy among the OECD nations. Freight trucks are projected to be the fastest growing mode of travel in the United States, with vehicle miles traveled by freight trucks increasing at an average rate of 2.2 percent per year from 2004 to 2030. U.S. air travel is projected to increase by an average of 1.7 percent per year over the period.

• Income growth and stable fuel prices are expected to continue the demand for larger, more powerful vehicles in the United States; however, advanced technologies and materials are expected to provide increased performance and size while improving new vehicle fuel economy. U.S. fuel economy standards for cars are assumed to remain at the current (2004) level of 27.5 miles per gallon through 2030. [This estimate has proved to be false.]

- The projected increase in transportation fuel in Mexico is based on expected growth in trade with the United States and overall improvement in the country's standard of living.

- Transportation energy demand in OECD Europe is projected to increase by only 0.2 percent per year. The transportation share of total energy use is projected to decline slightly. Low population growth, high taxes on transportation fuels, and environmental policies to discourage growth in transportation energy use are expected to slow the growth of transportation demand. [Western European countries are the only economies seriously trying to address climate change.]

- Historically, growth in transportation activity has been tied to income growth, indicating a strong relationship between per-capita GDP and passenger car travel per capita in countries with developing economies. In many countries of OECD D Asia, the availability of financing and an increase in the debt tolerance of middle class families are contributing to increased vehicle purchases.

- The transportation sector is projected to account for nearly 60 percent of the total increase in liquids use in non-OECD countries from 2004 to 2030. The growth in transportation energy use is expected to be led by greater demand for aviation fuel. Expanding ownership of private automobiles and an increasing role of trucking in freight transportation also play a significant role in the expected increase in energy demand.

- China's energy use for transportation is projected to grow by an average of 4.9 percent per year. Virtually all the growth in transportation energy consumption in China is projected to be in the form of liquids, mostly petroleum-based. As the country's economy expands, its energy use for air travel is expected to grow more rapidly than energy use for road transport. Personal travel in China has soared in the past two decades, with passenger miles traveled increasing fivefold. Still, in 2005 there were 4.5 million automobiles in China, as compared with 130.8 million automobiles in the United States.

- After China, India is expected to experience the fastest expansion in transportation sector energy use in the world, more than doubling transportation energy use between 2004 and 2030.

- The Middle East has a relatively small population and is not a major energy consumer but rather an exporter; however, rapid population growth in the region is expected to result in increased demand for transportation. The region's energy demand for transportation is projected to grow from 4.5 quadrillion BTU in 2004 to 9.0 quadrillion BTU in 2030.

- Residential sector energy use in the OECD countries accounts for about 60 percent of the world's residential delivered energy use, although the OECD nations account for only 18 percent of the world's population.

- China's strong growth in consumption helped to support high world oil prices in 2005 and 2006.

- Buses and two- and three-wheeled vehicles, which accounted for 42 percent of road energy use in China in 2004, are projected to decline to a 26-percent share in 2030, while the share represented by cars and light trucks increases from 18 percent in 2004 to 33 percent in 2030.

- Worldwide, the projected increase in residential electricity demand accounts for nearly 60 percent of the growth in overall residential energy demand from 2004 through 2030. By 2025, electricity overtakes natural gas as the world's largest source of energy for household use.

- The United States and OECD Europe together consumed nearly one-half (49 percent) of the world's delivered residential energy in 2004. In China and India, population growth, rising income levels and urbanization are expected to produce large increases in demand for residential energy services. Households in the non-OECD nations are projected to consume about 10 percent more energy than households in the OECD nations in 2030, as their economies continue to grow rapidly.

- In many non-OECD countries today, households still use traditional, non-marketed energy sources, including wood and waste, for heating and cooking. Regional economic development should displace some of that use as incomes rise and marketed fuels, such as propane and electricity, become more widely accessible.

- The need for services (health, education, financial, government) increases as populations increase. The degree to which these additional needs are met depends in large measure on economic resources — and economic growth. Higher levels of economic activity and disposable income lead to increased demand for hotels and restaurants to meet business and leisure requirements.

- The United States remains the largest consumer of commercial delivered energy in the OECD, accounting for one-half of the 24.6 quadrillion BTU of commercial energy use in the OECD as a whole in 2030.

- Coal remains an economically attractive choice for commercial water heating, space heating, and cooking in non-OECD countries in the projections, especially in China and India, which together account for around 80 percent of non-OECD commercial coal use from 2004 through 2030.

- The industrial sector is the largest of the end-use sectors, consuming more than 50 percent of the delivered energy worldwide in 2004. Worldwide, energy consumption in the industrial sector is projected to increase by an average of 1.8 percent per year from 2004 through 2030, as compared with 1.0-percent average annual growth in the global population. Industrial energy consumption is expected to increase in all countries and regions.

- In the United States, the manufacturing share of total economic output has declined steadily over the past two decades.
- Renewables remain a minor energy source for the industrial sector.
- The economies of many of the non-OECD countries and regions have growing energy-intensive, heavy manufacturing sectors.
- In the non-OECD economies' industrial sector, oil, coal, and natural gas were the most heavily used fuels in 2004, and they are projected to remain so in 2030.
- The continued importance of coal in the non-OECD industrial sector is largely attributable to China, which accounts for 70 percent of industrial coal use in the non-OECD economies in 2030.
- The use of petroleum and other liquid fuels is expected to increase globally by 35 million barrels per day by 2030. World liquid consumption was 83 million barrels per day in 2004 and is expected to increase to 118 million barrels per day in 2030.
- OPEC producers are expected to provide more than two-thirds of additional global oil production in 2030 — 23 million barrels per day.
- World liquids consumption increases to 118 million barrels per day in 2030, as the world continues to experience strong economic growth. The industrial sector accounts for a 27-percent share of the projected increase, mostly for use in chemical and petrochemical processes.
- The largest increases in consumption of oil between 2004 and 2030 are projected for North America and non-OECD Asia, at 7 and 15 million barrels per day, respectively.

The report predicts that world oil prices will decline from $68 per barrel in 2006 to $49 per barrel in 2014, then rise to $59 per barrel in 2030 ($95 per barrel on a nominal basis). In the low price case, world oil prices are projected to be $36 per barrel in 2030 ($58 per barrel on a nominal basis). In the high price case, oil prices are projected to be $100 per barrel in 2030 ($157 per barrel on a nominal basis). The report notes a "substantial range of uncertainty in the world's future oil markets."

- Increases in oil production are expected for both OPEC and non-OPEC producers; however, 65 percent of the total increase is expected to come from OPEC areas. In 2004, OPEC produced 41 percent of the world's liquids supply.
- Conflicts and social unrest could interrupt production and make investments more risky.
- After 2015, the reference case assumes that production decisions are made primarily on economic grounds, based on assessments of the resource base, with less weight placed on current political conditions.
- "The investment that several OPEC members (notably, Saudi Arabia and Angola) currently are making to expand their oil production capacity is ex-

pected to more than offset the slower expansion of non-OPEC supply projected in this year's outlook." North Sea production is projected to decline more rapidly than in last year's outlook.

- In Iran and Iraq, political developments are assumed to keep production levels fairly flat after 2015, when investment and production are projected to grow strongly through 2030.

- The report assumes a "business-as-usual" oil market environment. It does not consider disruptions in oil production for any reason such as war, terrorist activity, weather, or geopolitics.

- Most unconventional liquids are not economically competitive.

- The Caspian Basin output more than doubles from the 2004 level to 4.3 million barrels per day in 2015 in the reference case and increases steadily thereafter. Current uncertainty about export routes from the Caspian Basin region is "assumed to be resolved."

- In East Africa, Sudan is expected to produce significant volumes of oil by the end of this decade, with the potential to exceed 700,000 barrels per day in 2010.

- "Substantial exploration activity" is predicted in several West African nations.

- In North America, U.S. output that rises to 10.1 million barrels per day in 2020 and remains fairly flat through the end of the projection period is expected to be supplemented by significant production increases in Canada. The U.S. is relying heavily on unconventional output from Canadian oil sands projects.

- The IEA had to revise oil production estimates downward in both Mexico and the North Sea in the 2007 report. In the U.S. Gulf of Mexico, drilling depths routinely exceed 6,500 feet and can be more than 9,800 feet.

The report repeatedly alludes to U.S. frustration with Pemex, the Mexican national oil company. A clause in the Mexican constitution bars foreign investment in the Mexican oil industry. The report states: "Although the clause has allowed Pemex to maintain ownership of all its oil reserves, it also has prevented it from benefiting from technological advances that have allowed other national and major independent oil companies to improve their production opportunities." Also, "By some estimates, Pemex may need to invest as much as $32 billion annually in exploration and development to prevent a sharp decline in oil production." The report is critical of the Mexican Congress which redirects company profits to support government programs.

- Amidst enormous uncertainty, Iraq's role in OPEC in the next several years will "be of particular interest." Iraq's oil production capacity in 2007 is assumed to be 2.0 million barrels per day. Iraq has indicated a desire to expand production aggressively, to more than 6 million barrels per day, once the security and political situation in the country has stabilized. Preliminary discussions of exploration projects have already been held with a number of potential outside investors. In the reference case, Iraq's oil production is projected to reach 3.3 million barrels per day in 2015 and 5.3 million barrels per day in 2030.

• Oil production in Iran is projected to increase only slightly in the early years of the reference case, from 4.1 million barrels per day in 2004 to 4.3 million barrels per day in 2015, despite the country's sizable resource base. In the long run, Iran's oil production is projected to reach 5 million barrels per day in 2030.

• Geopolitical issues in a number of the OPEC countries, including Iraq, Iran, Venezuela, and Nigeria make it difficult to estimate future production levels.

• Oil prices fall as non-OPEC production expands, and OPEC producers must increase production to meet their revenue requirements. As a result, OPEC's options for influencing the market are limited.

• Historically, estimates of world oil reserves have generally trended upward. As of January 1, 2007, proved world oil reserves, as reported by *Oil & Gas Journal*, were estimated at 1,317 billion barrels — 24 billion barrels (about 2 percent) higher than the estimate for 2006. Taken together, the reserve increases and production imply that 54 billion barrels of reserve discoveries and growth occurred during 2006, or an increase of about 4 percent.

• Iran ranks third in world oil reserves with 136.3 billion barrels; Iraq ranks fourth with 115.0. Iranian oil reserves have increased by 52 percent since 2000.

• Fifty-six percent of the world's total proved oil reserves are located in the Middle East. Among the top 20 reserve holders in 2007, 11 are OPEC member countries that, together, account for 65 percent of the world's total reserves. The largest declines in oil reserves between 2000 and 2007 were reported in Mexico, China, Norway, Australia, and the United Kingdom.

The most common measure of the adequacy of proved reserves relative to annual production is the reserve-to-production (r/p) ratio, which describes the number of years of remaining production from current proved reserves at current production rates.

• Iraq leads the pack with 168 years; Iran has a ratio of 83 years. The United States has a ratio of 11 years; Canada is 10 years. Saudi Arabia's ratio is 75 years.

• Natural gas consumption in the non-OECD countries grows more than twice as fast as in the OECD countries. Production increases in the non-OECD region account for more than 90 percent of the growth in world production from 2004 to 2030.

- By energy source, the projected increase in natural gas consumption is second only to coal. Natural gas remains a key fuel in the electric power and industrial sectors. Natural gas burns more cleanly that coal or petroleum products.
- In 2030, more than one third of the natural gas consumed in OECD countries is projected to come from other parts of the world, up from 22 percent in 2004.
- Almost three-quarters of the world's natural gas reserves are located in the Middle East and Eurasia. Russia, Iran, and Qatar combined accounted for about 58 percent of the world's natural gas reserves as of January 1, 2007.
- The Middle East's reserves-to-production ratio exceeds 100 years. The worldwide ratio is 65 years.
- An estimated 3,000 trillion cubic feet of natural gas is in "stranded" reserves, usually located too far away from pipeline infrastructure or population centers for its transportation to be economical.
- Russia has 27.2 percent of world's natural gas reserves, followed by Iran with 15.8 percent. In comparison, the U.S. has 3.3 percent.
- In 2030, the export share of natural gas production from non-OECD Asia is projected to fall to 10 percent, as domestic consumption takes precedence over exports.
- Alaska's natural gas production accounts for all of the projected growth in domestic U.S. conventional natural gas production from 2004 to 2030.
- A large portion of North America's remaining technically recoverable resource base of natural gas consists of unconventional sources, which include tight sands, shale, and coal bed methane. With most of the large onshore conventional fields in the United States already having been discovered, the United States, like Canada, must look to these costlier sources of supply to make up for declines in conventional production.
- The significant growth in U.S. liquefied natural gas imports is indicative of the country's growing dependence on imports and the increasing globalization of natural gas markets.

Investment in Australia's natural gas sector projects has been helped by the country's reputation as a stable political environment that takes no state equity in reserves or liquefied natural gas assets. Even in Australia, however, state involvement has a bearing on project economics. The Gorgon LNG project to develop reserves off Australia's northwest coast faces not only stringent environmental requirements but also, in an agreement with the Western Australian state government, a requirement that the project must allocate 15 percent of the Gorgon reserves for domestic consumption. The Western Australian government is now looking at options for applying domestic reservation requirements to all future liquefaction projects that would process offshore gas in Western Australia.

Increasing state involvement in the upstream natural gas activities of several large reserve holders throughout the world is threatening to delay or discourage investments in new production and export capacity. In May 2006, Bolivia nationalized its energy resources, prompting investors to suspend further investment.

It seems that everyone from Venezuela to Russia to the United States views natural gas and petroleum as strategic assets, and the temptation to export for cash has to be weighed against the need to hold onto reserves for the future.

• As the largest consumer, the United States accounted for more than 80 percent of the 27.6 trillion cubic feet of natural gas consumed in North America in 2004.

• After 2015, higher natural gas prices in the United States, along with tax incentives for clean coal technologies, are expected to discourage the construction of new natural gas-fired plants in favor of coal-fired plants, leading to a decline in the natural gas share to 16 percent and an increase in the coal share to 57 percent in 2030. ("Clean coal" technologies are very much in dispute.)

• Natural gas-fired generation is less carbon-intensive than oil- or coal-fired generation and is expected to remain more cost-competitive than renewable energy, making natural gas the fuel of choice for new generating capacity in OECD Europe.

• Although Australia has not ratified the Kyoto Protocol, several of the government's environmental policies have been put in place to help stimulate increases in natural gas use for electric power generation and to moderate growth in the use of coal, of which Australia has large reserves.

• In both China and India, natural gas currently is a minor fuel in the overall energy mix; however, both countries are rapidly expanding infrastructure to serve demand.

• Natural gas consumption grows at average annual rates of 2.5 percent in the Middle East and 3.3 percent in Africa from 2004 to 2030. Natural gas consumption in the Middle East almost doubles over the projection period and consumption in Africa more than doubles. After 2015, production increases in the two regions are expected to be directed more toward domestic consumption.

• World coal consumption increases by 74 percent from 2004 to 2030, international coal trade increases by 44 percent from 2005 to 2030, and coal's share of world energy consumption increases from 26 percent in 2004 to 28 percent in 2030.

• Although coal currently is the second-largest fuel source of energy-related carbon dioxide emissions (behind oil), accounting for 39 percent of the world total in 2004, it is projected to become the largest source by 2010. Carbon dioxide emissions per unit of energy output are higher for coal than for oil or natural gas. In 2030, coal's share of energy-related carbon dioxide emissions is projected to be 43 percent, compared with 36 percent for oil and 21 percent for natural gas.

• Total recoverable reserves of coal around the world are estimated at 998 billion tons — reflecting a current reserves-to-production ratio of 164. [In other words, coal can be burned for the next 164 years.]

• Sixty-seven percent of the world's recoverable coal reserves are located in four countries: the United States (27 percent), Russia (17 percent), China (13 percent), and India (10 percent). In 2004, these four countries, taken together, accounted for 66 percent of total world coal production. From 2004 to 2030, coal production in China, the United States, and India, driven by growing coal consumption, is projected to increase by 50.4 quadrillion BTU, 11.1 quadrillion BTU, and 5.7 quadrillion BTU, respectively. It is assumed that most of the demand for coal in China, the United States, and India will continue to be met by domestic production. [In other words, the world's three largest economies will rely on their considerable coal reserves to fuel their countries.] The United States has a total of 267.6 billion short tons of coal, by far the most in the world. China has 126.2 and India 101.9.

• Much of the increase in coal consumption projected for the OECD countries from 2004 to 2030 is the result of expected strong growth in U.S. coal demand, under the assumption that existing laws and polices remain in effect indefinitely. [This is a grandiose assumption.]

• In 2004, the United States consumed 22.6 quadrillion BTU of energy from coal, accounting for 94 percent of total coal consumption in North America and 48 percent of the OECD total. U.S. coal consumption rises to 34.1 quadrillion BTU in 2030 in the reference case. The United States has substantial coal reserves and has come to rely heavily on coal for electricity generation, a trend that continues in the projections. Much of the projected growth in U.S. coal consumption occurs after 2015. After 2015, rising natural gas prices gradually tilt economic decisions toward new coal-fired power plants.

• In Canada, the Ontario government is moving ahead to shut down coal-fired plants. This decision was based on the premise that the adverse health and environmental impacts of the plants' operation were unacceptable.

• In contrast to the United States, coal consumption in OECD Europe declines by 1.7 quadrillion BTU (13 percent) from 2004 to 2030 in the reference case.

• A key incentive for the new coal builds in Germany is a provision guaranteeing carbon dioxide emission rights for the new capacity during the first 14 years of its operation. This provision has been adamantly opposed by environmentalists. There has been continuing pressure on members of the European Union to reduce subsidies that support domestic production of hard coal.

- In 2004, Australia was the world's leading coal exporter while Japan and South Korea were the world's leading importers. With substantial coal reserves, the Australia/New Zealand region continues to rely heavily on coal for electricity generation; however, coal's share of total generation in the region is projected to decline gradually, as more natural gas is used for power generation.

- Led by strong economic growth and rising demand for energy in China and India, non-OECD coal consumption is projected to rise to 139.8 quadrillion BTU in 2030, more than double the quantity consumed in 2004. The increase of 71.9 quadrillion BTU represents 85 percent of the projected increase in total world coal consumption.

- China and India together account for 72 percent of the projected increase in world coal consumption from 2004 to 2030. Much of the increase in their demand for energy is expected to be met by coal. China was the world's leading producer of both steel and pig iron in 2004. Coal remains the primary source of energy in China's industrial sector, primarily because the country has only limited reserves of oil and natural gas. Although China's government and industry have proposed to build as much as 1 million barrels of daily coal-to-liquids capacity by 2020, considerable uncertainty and risks are associated with the emergence of such a massive coal-to-liquids industry, including potential strains on water resources and the general financial risks associated with the technological uncertainties and huge capital investments.

- Nearly 70 percent of the growth in India's coal consumption is expected to be in the electric power sector and most of the remainder in the industrial sector. India's government is also pursuing the development of between five and seven large coal-fired power projects.

- Significant growth in coal consumption is expected in Taiwan, Vietnam, Indonesia, and Malaysia, where considerable amounts of new coal-fired generating capacity are either planned or under construction.

- Non-OECD Europe and Eurasia have substantial coal reserves: Russia alone has an estimated 173 billion tons of recoverable reserves (17 percent of the world total), and the other countries in the region have an additional 100 billion tons (10 percent of the world total). In 2030, Russia's coal use is projected to total 6.1 quadrillion BTU. Natural gas is expected to be the most economical option for new generating capacity in Russia. The government maintains that fossil fuel-fired plants will continue in their role as the primary source for electric power generation through 2020. Siberia is a coal-rich region. Plans for both new coal-fired capacity and refurbishment of existing capacity in a number of countries, including Bosnia and Herzegovina, Serbia and Montenegro, Bulgaria, Romania, and Ukraine, are a significant indication that coal will continue to be an important source of energy for the region.

• Recent power shortages and the general lack of spare generating capacity in southern Africa have led to increased interest in new coal-fired projects in South Africa, Mozambique, Zimbabwe, Tanzania, and Botswana.

• Brazil, with the world's eighth-largest steel industry in 2004, accounted for 56 percent of the region's coal demand. Chile, Colombia, Puerto Rico, Peru, and Argentina accounted for most of the remainder.

• Israel accounted for 86 percent of coal consumption in the Middle East in 2004. The region's consumption increases only slightly in the projections.

• Australia and Indonesia are geographically well suited to continue as the leading suppliers of internationally-traded coal (particularly to Asia) until 2030. For coking coal, Australia, Canada, and the United States continue to be ranked among the top three exporters over the projection period. Australia is projected to dominate future international coal trade.

• China and Vietnam are examples of countries that have the potential to export more coal but are focused instead on meeting domestic coal demand. China has the potential to influence the market either as an importer or an exporter. Most of the coal consumed in China is expected to come from its own coal mines.

• In India, demand for coal imports in 2030 is projected to be almost double the demand in 2005, as the country continues to encounter problems with coal production and transportation within its borders. Japan, lacking coal resources of its own, is expected to remain the world's largest importer of coal in 2030.

• For most European countries with increasing emphasis on natural gas in the power sector, coal becomes a less significant component of the fuel mix for electricity generation. Imports may be generated because of a phase-out of European mining subsidies and higher demand for lower sulfur coal.

• The United States is projected to import 2.1 quadrillion BTU of coal in 2030, 1.3 quadrillion BTU more than in 2005. Although still a small share of U.S. consumption, at 6.1 percent, that would represent a shift for the United States from being a net exporter to being a net importer.

• With declining productivity and mining difficulties in Central Appalachia, and with rising demand for coal in the Southeast, imports are expected to become increasingly competitive with domestic U.S. coal production. Already, plans are being made to expand U.S. ports to accommodate coal imports. For example, Kinder Morgan Energy Partners LP is adding 9 million tons of coal import capacity at its Virginia port facilities in 2008. South America is expected to be an important source of coal imports to the United States.

• World electricity generation is expected to nearly double from 2004 to 2030. In 2030, generation in the non-OECD countries is projected to exceed generation in the OECD countries by 30 percent. Overall global demand advances strongly during this period. The most rapid growth in total world demand for electricity is projected for the buildings (residential and commercial) sectors. Growth in non-OECD countries increases demand for office space, hospitals, hotels, and other institutions or organizations. Coal has continued to be the fuel most widely used for electricity generation.

• High world oil prices — which have been trending upward since 2003 — have reinforced coal's important role as an energy source for electricity generation. Projections of future coal use are particularly sensitive to assumptions about future policies that might be adopted to mitigate greenhouse gas emissions. [While world oil prices have dropped after a huge spike, coal will play a vital role in electricity generation, as it does today.]

• Coal continues to provide the largest share of the energy used for electric power production. In 2004, coal-fired generation accounted for 41 percent of the world electricity supply; in 2030, its share is projected to be 45 percent. Sustained high prices for oil and natural gas make coal-fired generation more attractive economically, particularly in nations that are rich in coal resources, which include China, India, and the United States.

• The total amount of electricity generated from natural gas continues to be only about one-half the total for coal, even in 2030. Natural gas burns more cleanly than coal or petroleum products, and as more governments begin implementing national or regional plans to reduce carbon dioxide emissions, they may encourage the use of natural gas to displace oil and coal.

• Only the Middle East region, with its ample oil reserves and a current one-third share of total electricity generation fueled by oil, is projected to continue relying heavily on oil to meet its electricity needs.

• Issues related to nuclear plant safety, radioactive waste disposal, and the proliferation of nuclear weapons, which continue to raise public concerns in many countries, may hinder the development of new nuclear power reactors. But only OECD Europe is projected to see a decline in nuclear power generation after 2010.

• Hydroelectric and other renewable energy resources are projected to increase at an average annual rate of 1.7 percent from 2004 to 2030. Renewable energy is attractive for environmental reasons. However, renewables cannot compete economically with fossil fuels. The renewable share of world electricity generation falls slightly from 19 percent in 2004 to 16 percent in 2030, as growth in the consumption of both coal and natural gas in the electricity generation sector worldwide exceeds the growth in renewable energy consumption. The capital costs of new power plants using renewable fuels remain relatively high in comparison with those for plants fired with coal and natural gas.

- Electricity demand in OECD North America is projected to grow by an annual average of 1.5 percent from 2004 to 2030, which is less than one-half the projected rates of increase in China and in India. Still, North America is projected to account for 23 percent of the world's electric power generation in 2030. The United States is the largest consumer of electricity in North America and is projected to remain in that position through 2030.

- There are large differences in the mix of energy sources used to generate electricity in the United States, Canada, and Mexico, and those differences are likely to become more pronounced in the future. In the United States, coal accounted for 52 percent of the 2004 total; but in Canada, renewable energy sources (predominantly hydroelectricity) provided 60 percent of the nation's electricity generation in 2004. Most of Mexico's electricity generation is fueled by petroleum-based liquids and natural gas. In the reference case projections for 2030, U.S. reliance on coal is even greater than it was in 2004. Oil-fired generation is projected to decline in Canada. The Province of Ontario had announced plans to close all its coal-fired plants by the end of 2007, but that date has been pushed back to 2011. Given Canada's experience with hydropower and the commitments for construction, new hydroelectric capacity accounts for more than one-half of additional renewable capacity projected to be added in Canada between 2004 and 2030. Provincial governments also offer incentives for renewables like wind power for electricity generation.

- In OECD Europe, natural gas is expected to be by far the fastest growing fuel for electricity generation, increasing at an average rate of 3.3 percent per year from 2004 to 2030, while high world oil prices and environmental concerns lead to decreases in the use of petroleum and coal. At present, 7 of the world's 10 largest markets for wind-powered electricity generation are in Europe, and the 27-member European Union accounted for 65 percent of the world's total installed wind capacity as of the end of 2006. The role of wind power in meeting OECD Europe's electricity demand is likely to grow in the future.

- In Australia and New Zealand, natural gas-fired generation is expected to grow strongly in the region, reducing coal's share in 2030. [It is interesting to note that Australia is the world leader in coal exports but plans to rely more on natural gas for domestic use for environmental reasons. Australia seems willing to sell, but not to use, coal.]

- China already is the world's largest coal consumer, followed by the United States, and India ranks third. In 2004, the combined coal use of China and India was equal to that of the entire OECD region, and in 2030 it is projected to exceed the OECD total by more than 85 percent. In non-OECD Asia, the renewable share of total generation declines as the shares of fossil fuels and nuclear power grow more strongly in the region. The renewable share of generation falls from 16 percent in 2004 to 9 percent in 2030.

• Population and income growth in the Middle East are expected to result in growing demand for electric power in the future. Petroleum is a valuable export commodity for many nations of the Middle East, and there is increasing interest in the use of domestic natural gas for electricity generation in order to make more oil assets available for export. Because there is little incentive for countries in the Middle East to increase their use of renewable energy sources, renewables are projected to account for a modest 2 percent of the region's total electricity generation throughout the projection period.

• In Central and South America, heavy reliance on hydroelectricity has been problematic in times of drought. In Brazil, hydropower is projected to remain the predominant source of electricity through 2030. Still, hydropower and other renewable energy sources in their combined fuel use for electricity generation are projected to be 47 percent in Central and South America.

• World carbon dioxide emissions are projected to rise from 26.9 billion metric tons in 2004 to 42.9 in 2030 — that is a 16 billion metric ton increase.

• In 2004, non-OECD emissions of carbon dioxide were greater that OECD emissions for the first time. In 2030, carbon dioxide emissions from the non-OECD countries are projected to exceed those from the OECD countries by 57 percent.

• Coal's share of carbon dioxide emissions in 2004 was the same as its share in 1990, at 39 percent; however, its share is projected to increase to 43 percent in 2030. Coal is the most carbon-intensive of the fossil fuels, and it is the fastest-growing energy source in the reference case projection.

• In 1990, China and India combined for 13 percent of world emissions, but by 2004 that share had risen to 22 percent — largely because of a strong increase in coal use in these two countries. This trend is predicted to continue; and by 2030, carbon dioxide emissions from China and India combined are projected to account for 31 percent of total world emissions, with China alone responsible for 26 percent of the world total. As both economies expand, coal will become a greater part of the world energy mix and play a correspondingly larger role in the composition of world carbon dioxide emissions.

In an amazing example of understatement, the report states: "There are some signs that concerns about global climate change are beginning to affect the world fuel mix."

• The highest growth rate in the non-OECD regions is projected for China, at 3.4 percent annually from 2004 to 2030, reflecting the country's continued heavy reliance on fossil fuels, especially coal, over the projection period. China's energy-related emissions of carbon dioxide are projected to exceed U.S. emissions by about 5 percent in 2010 and by 41 percent in 2030.

• The United States is expected to remain the largest source of petroleum-related carbon dioxide emissions throughout the projection period, with projected emissions of 3.3 billion metric tons in 2030 — still 66 percent above the corresponding projection for China. The growth in U.S. emissions is projected to average 0.6 percent per year, but the projected level of 1.4 billion metric tons in 2030 is more than triple the projection for China.

• Despite the above-mentioned statistics, the report concluded: "In all countries and regions, carbon dioxide intensity — expressed in emissions per unit of economic output — are projected to improve (decline) over the projection period as the world economy moves into a post-industrial phase. In 2004, estimated carbon dioxide intensity was 470 metric tons per million dollars of GDP in the OECD region and 516 metric tons per million dollars in the non-OECD region." The U.S. carbon dioxide intensity in 2030 is projected to be 353 metric tons per million dollars of GDP.

• Another measure of carbon dioxide intensity is emissions per person. Carbon dioxide emissions per capita in the OECD region are significantly higher than in the non-OECD region. If non-OECD countries consumed as much energy per capita as the OECD countries, the projection for world carbon dioxide emissions in 2030 would be much larger, because the non-OECD countries would consume about 3.5 times more energy than the current reference case estimate of 404 quadrillion BTU.

• Russia has the highest projected increase in carbon dioxide emissions per capita in the reference case; India and Africa have the lowest.

• In the United States, emissions per capita are projected to rise from 20 metric tons in 2004 to 22 metric tons in 2030. This compares with 19 metric tons in Canada.

There is a strong correlation between income and emissions per capita.

• The projected economic growth in India is less carbon-intensive than in other countries, as it moves more toward service industries rather than energy-intensive manufacturing. Per capita GDP in India is projected to grow by 4.5 percent per year from 2004 to 2030, while its carbon dioxide emissions per capita are projected to increase by only 1.5 percent per year.

• Some other figures are worth examining in the Energy Information Report. The general impression is that the juggernauts of China and India are vying to dethrone the United States in the arena of world economics. The gross domestic product (GDP) of the United States is projected to be 22,494 billion dollars in 2030. [The actual figure is unknowable because of the recent global recession.] The combined GDP of China and India is projected to be 11,270 billion dollars, slightly more than half of U.S. GDP. The U.S. population is expected to be about 400 million in 2030. The combined populations of China and India are projected to be 3 billion, over seven times more populated than the United States.

• World oil consumption increases from 82.5 million barrels per day to 117.6 million barrels per day in 2030 — an increase of 35.1 million barrels per day.

• World natural gas consumption increases from 99.6 trillion cubic feet to 163.2 trillion cubic feet in 2030 — an increase of 63.6 trillion cubic feet. The United States will consume 26.1 trillion cubic feet in 2030; China will consume 7 and India 3.9.

• World consumption of coal goes from 114.5 quadrillion BTUs [British Thermal Units] to 199.1 in 2030 — an increase of 84.6 quadrillion BTUs.

• World renewables (which is mostly hydroelectricity) increases from 33.2 quadrillion BTUs to 53.5 quadrillion BTUs by 2030.

• World carbon dioxide emissions increase from 26,922 million metric tons to 42,880 million metric tons in 2030 — an increase of 15,958 million metric tons. The United States will be second only to China in carbon dioxide emissions with 7,950 million metric tons. China will emit 11,239 million metric tons. By contrast, OECD Europe will emit 4,684 million metric tons. India will emit 2,156 metric tons.

• World carbon dioxide emissions from oil and other liquids increases 4,559 million metric tons by 2030 for a total of 15,411 million metric tons.

• World carbon dioxide emissions from natural gas increase by 3,547 million metric tons by 2030 for a total of 8,988 million metric tons.

• World carbon dioxide emissions from coal increases 7,849 million metric tons by 2030 for a total of 18,466 million metric tons.

• By 2030, when the world population is predicted to have increased by 1,915 million people for a total of over 8.2 billion people, the OECD countries will consume 326 BTU of energy by 2030. Non-OECD countries will consume 252.

• India will consume the most nuclear energy, followed by China and Russia. The only decrease in nuclear energy will be in OECD Europe.

• The much-heralded renewable sector is barely a blip in the total world delivered energy consumption by the end-use sector and fuel. Renewables are projected to be 4.3 quadrillion BTU in 2030 in all end-use sectors, up from 2.9 in 2004.

• By far the most used renewable resource for electricity generation is hydropower, primarily in South America, Central America, and Canada. The global use of renewables in OECD countries is projected to rise 1.2 percent from 2004 to 2030. The consumption of renewable energy in the U.S. is expected to increase by 1.4 percent by 2030, for a renewable consumption rate of 8.7 quadrillion BTU. Oil and other liquids consumption, by comparison, will be 52.1 quadrillion BTU.

• Renewables consumption in OECD Europe is projected to rise only 0.9 percent by 2030 from 6.3 quadrillion BTU to 8 quadrillion BTU. Europe is expected to become more reliant on natural gas. Natural gas is cleaner than coal but still emits carbon dioxide. Renewables in Japan will only increase 1 percent to 1.5 quadrillion BTU by 2030. Renewables in Australia and New Zealand will increase by 1.3 percent, from a miniscule 0.5 quadrillion BTU to 0.7 in 2030. As in Europe, Australia seems to be moving more toward natural gas even though it is the world's leading exporter of coal. African renewables consumption is projected to rise only 1.5 percent, from 0.9 to 1.3 quadrillion BTU.

• OPEC's share of the world's liquids production is expected to rise from 42 percent in 2005 to 47 percent in 2030. The Persian Gulf's share of world production is projected to rise from 29 percent to 32 percent by 2030.

• Electricity generation using renewables is expected to rise a paltry 0.6 percent in the United States from 2004 to 2030. In 2004, the U.S. had 122 gigawatts of hydroelectric and other renewable generating capacity. That capacity is project to rise to 142 gig watts by 2030. Brazil will lead the world in renewables with a 3.3 percent increase due in large part to hydroelectricity projects.

• The OECD countries make up 18 percent of the world population; non-OECD countries constituted the other 82 percent. The OECD countries are the United States, Canada, Mexico, Austria, Belgium, Czech Republic, Denmark, Finland, France, Germany, Greece, Hungary, Iceland, Ireland, Italy, Luxembourg, the Netherlands, Norway, Poland, Portugal, Slovakia, Spain, Sweden, Switzerland, Turkey, the United Kingdom, Japan, South Korea, Australia, and New Zealand.

• The International Energy Agency in Paris recently released their own "World Energy Outlook 2008 Fact Sheet: Global Energy Trends"[1] to address the question: "Where are we headed without new policies and what does it mean?" Some highlights of the report, which take the global economic crisis into consideration, are as follows:

• The impact of the credit crisis on world economic growth prospects, higher energy prices and some notable new policy initiatives have left their mark on the *World Energy Outlook 2008*. Energy use grows more slowly to 2030 than projected last year, but the overall trends are broadly unchanged: persistent dominance of fossil fuels — oil, gas, and coal — in the energy mix; a rising share of emerging economies in global energy consumption; an increase in the consuming countries' reliance on imports of oil and gas; and an inexorable rise in global CO_2 emissions.

• World primary energy demand expands by 45 percent between 2006 and 2030 — an average rate of growth of 1.6 percent per year. Fossil fuels account for 80 percent of the world's primary energy mix in 2030 — down only slight-

---

1 *"World Energy Outlook Fact Sheet: Global Energy Trends,"* International Energy Agency, Paris, France, 2008.

ly on today. Oil remains the dominant fuel, though demand for coal rises more than any other fuel.

- China and India account for over half of incremental energy demand to 2030. The Middle East emerges as a major new demand center, contributing a further 11 percent to incremental world demand. Collectively, non-OECD countries account for 87 percent of the increase, their share of world primary energy demand rising from 51 percent to 62 percent.

- M]uch of the additional imports have to transit vulnerable maritime routes.

- Cumulative investment in energy-supply infrastructure amounts to $26.3 trillion to 2030.

- Trends highlight the extent of the challenge of securing the supply of reliable and affordable energy and effecting a rapid transition to a low-carbon, efficient and environmentally benign energy system. The Reference Scenario, characterized by rising energy prices, increased import dependence and rising greenhouse-gas emissions, is unsustainable — environmentally, economically, and socially. Achieving a more secure, low-carbon energy system calls for radical action by governments at national and local levels, and through participation in coordinated international mechanisms.

- Assuming no changes in government policies, world oil demand is set to continue to expand through to 2030 albeit more slowly with rises by 1 percent per year on average, from 85 million barrels per day in 2007 to 106 million barrels per day in 2030.

- India sees the fastest growth (3.9 percent per year), followed by China (3.5 percent) which is significantly lower than in the past. The share of OECD countries (North America, Europe, and the Pacific) in global oil demand drops from 57 percent in 2007 to 43 percent in 2030.

- The majority of oil consumers worldwide do not pay prices that fully reflect international market levels.

- The sheer growth of the car fleet — from an estimated 650 million in 2005 to about 1.4 billion by 2030 — is expected to continue to push up total oil use for transport purposes. There is not expected to be any major shift away from conventionally-fuelled vehicles before 2030, though the penetration of hybrid-electric cars is projected to rise, reducing oil demand growth.

- As a share of world GDP at market exchange rates, oil spending soared from a little over 1 percent in 1999 to around 4 percent in 2007, contributing to the economic downturn experienced by most oil-consuming countries. That share is projected to stabilize at around 5 percent over much of the projection period. For non-OECD countries, the share averages 6 to 7 percent.

- Almost all the additional capacity from new oilfields is offset by declines in output at existing fields. The bulk of the net increase in total oil production comes from NGLs [natural gas liquids, or liquid natural gas] (driven by the relative rapid expansion in gas supply) and from non-conventional resources and technologies, notably Canadian oil sands.

• Saudi Arabia remains the world's largest producer throughout the projection period, its output climbing from 10.2 million barrels a day in 2007 to 15.6 million barrels a day in 2030.

• The world's endowment of oil is large enough to support the projected rise in output, but rising oilfield decline rates will push up investment needs.... Decline rates ... are set to accelerate in the long term in each major world region. The average observed decline rate worldwide is currently 6.7 percent for fields that have passed their production peak. This rate rises to 8.6 percent in 2030.

• Even if oil demand was to remain flat to 2030, 45 million barrels per day of gross capacity — roughly four times the current capacity of Saudi Arabia — would need to be built worldwide by 2030 just to offset the effect of oilfield decline.

• More capacity will need to be sanctioned within the next two years to avoid a fall in spare capacity towards the middle of the next decade and a possible supply crunch. In view of the current financial crisis, there are growing doubts about whether all of this capacity will be forthcoming.

• Rising global consumption of fossil fuels is still set to drive up greenhouse-gas emissions and global temperatures, resulting in potentially catastrophic and irreversible climate change. The projected rise in emissions in the Reference Scenario, in which no change in government policies is assumed, puts us on a course of doubling the concentration of those gases in the atmosphere to around 1,000 parts per million CO_2-equivalent by the end of this century. This would lead to an eventual global temperature increase of up to 6 degrees Celsius.

• Global energy-related CO_2 emissions are projected to rise from 28 gigatons (Gt) in 2006 to 41 Gt in 2030 — an increase of 45 percent. The 2030 figure is only 1 Gt lower than that projected in last year's *Outlook*, even though we assume slower world economic growth and higher energy prices. World greenhouse-gas emissions, including non-energy CO_2 and all other gases, are projected to grow from 44 Gt CO_2-eq in 2005 to 60 Gt CO_2-eq in 2030 — an increase of 35 percent.

• Three quarters of the projected increase in energy-related CO_2 emissions arises in China, India, and the Middle East, and 97 percent in non-OECD countries as a whole.... Only in Europe and Japan are emissions in 2030 lower than today. The bulk of the increase ... is expected to come from cities, their share rising from 71 percent in 2006 to 76 percent in 2030 as a result of urbanization. City residents tend to consume more energy than rural residents, so they emit more CO_2 per capita.

- The power generation and transport sectors contribute over 70 percent of the projected increase in world energy-related CO_2 emission to 2030.
- Three-quarters of the projected output of electricity worldwide in 2020 (and more than half in 2030) comes from power stations that are already operating today. The rate of capital-stock turnover is particularly slow in the power sector, where large up-front costs and long operating lifetimes mean that plants that have already been built are effectively "locked-in." As a result, even if all power plants built from now onwards were carbon-free, CO_2 emissions from the power sector would still be only 25 percent, or 4 Gt, lower in 2020 relative to the Reference Scenario.
- The share of low-carbon energy — hydropower, nuclear, biomass, renewable and fossil-fuel power plants with carbon capture and storage (CCS) — in the world primary energy mix increases from 19 percent in 2006 to 25 percent in 2030.
- The OECD countries alone cannot put the world onto the path to 450 parts per million trajectory, even if they were to reduce their emissions to zero.
- Oil and gas exports in the top-ten producing sub-Saharan African countries are set to grow steadily to 2030, providing the means for alleviating poverty and expanding energy access.

The U.S. Department of Energy, their government scientists, and companies in the U.S. energy sector have been hard at work with their efforts to deal with the impending climate change caused by the human overuse of fossil fuels which they have effectively denied until very recently. The U.S. Climate Change Science Program, which was the Bush Administration's answer to the Kyoto Protocol, issued a new report in October 2007 addressing climate change.[1] The report's authors spent little time on climate science because the problems have become widely known and those interested can find voluminous information on the topic elsewhere.

The scientists contributing to the report sounded almost irate, in an Orwellian sort of way. Sure, the United States energy sector has generated the overwhelming amount of carbon dioxide that has caused the global problems in the first place and at a hefty profit, but *what about us?* Rather than address the energy sector as the *driver* of global warming, they worried about how global warming will *drive* the energy sector and considered how they could make the best of a bad situation. *Don't we have our own problems? Doesn't anybody care that our oil platforms are being dismantled by this damn weather?* In other words, the report is concerned with the effects of climate change on the energy industry, not the effects of the energy industry on climate change.

---

1 *"Third U.S. Climate Change Science Program Report,"* U.S. Department of Energy, Washington, D.C., Oct. 18, 2007.

Fortunately for the energy industry, these scientists see the glass as half full. According the report, "the natural balance (is) slightly favoring net savings of delivered energy." Cutting through the lumbering prose, the situation is explained thusly: Increased temperature will cause more demand for cooling in the southern region and parts of the northern region that, in the past, have not found air conditioning necessary. On the up side, less heating will be required in the northern climes. This includes Canada, where some energy once used for heating could now be purloined by its neighbors to the south, especially hydroelectric power from Quebec. The improvements in energy efficiency in commercial and residential buildings will also decrease energy needs. But not all is well in the Land of Nod.

"Climate change would cause a significant increase in the demand for electricity in the United States costing billions of dollars," according to the report, but "it is not thought that industrial energy demand is particularly sensitive to climate change."

"Net effects of climate change in the United States on total energy demand are projected to be modest," the report continues.

However, it will become more challenging to pump out fossil fuels into the atmosphere for one significant reason — water availability.

"Fossil fuels place a high demand on the nation's water resources. Power plants rank only slightly behind irrigation in terms of freshwater withdrawal in the United States." Water is also required in the mining, processing, and transportation of coal to generate electricity, all of which can have direct impacts on water quality. The United States Geological Survey estimated in 2000 that the mining industry withdrew approximately 2 billion gallons of freshwater per day."[1]

The U.S. freshwater problems have already been documented in Chapter One of this book. According to another DOE-sanctioned publication, *International Energy Outlook 2007*, prepared by the Energy Information Administration, the energy sector will rely heavily on its ample coal reserves — coal being by far the dirtiest fuel of all — through 2030. China will also rely heavily on coal and has water problems of its own.

The U.S. General Accountability Office (GAO) reported in a 2003 study that 36 states anticipate water shortages in the next 10 years under *normal* conditions.[2] Power plants across the country are being denied permits because they would use too much water.

The United States is also banking heavily on oil shale deposits and oil sand fields in Canada. Both processes involve voluminous amounts of water. Another

---

1 Susan S. Hutson *et al.*, *"Estimated Use of Water in the United States in 2000,"* U.S. Geological Survey, released March 2004, revised April 2004, May 2004, and February 2005, Denver, Colo.

2 *"Freshwater Supply: States' Views of How Federal Agencies Could Help Them Meet the Challenges of Expected Shortages,"* U.S. General Accounting Office, July 2003.

strategy, coals-to-liquids operations, also requires significant quantities of water. Oil shale is located primarily in arid regions.

NASA scientist James Hansen, who consistently battled the Bush Administration over the climate change issues and the administration's propensity to ignore or downplay global warming, had once been a vocal proponent of carbon capture and sequestration from oil-fired coal plants. But Hansen came to believe that coal should simply be left in the ground. And the already-fouled climate may conspire to derail the effectiveness of this technology. Increased ambient temperatures can degrade pipeline system performance and, if practiced in the future, carbon sequestration.

Hansen testified recently before the Iowa Utilities Board to explain the impact of coal-fired power plants on global warming and explain what scientists know about climate change in general.[1]

> My aim is to present clear scientific evidence describing the impact that coal-fired power plants ... will have on the Earth's climate, and thus on the well-being of today's and future generations of people and on all creatures and species of creation,... Burning of fossil fuels, primarily coal, oil, and gas, increases the amount of carbon dioxide and other gases and particles in the air. These gases and particles affect the Earth's energy balance, changing both the amount of sunlight absorbed by the planet and the emission of heat (long wave or thermal radiation) to space. The net effect is a global warming that has become substantial during the past three decades.

> Global warming from continued burning of more and more fossil fuels poses clear dangers for the planet and for the planet's present and future inhabitants. Coal is the largest contributor to the human-made increase of CO_2 in the air. Saving the planet and creation surely requires phase-out of coal use.

The United States and Europe together are responsible for well over half of the increase from the pre-industrial CO_2 amount (280 parts per million) to the present day CO_2 amount (about 387 ppm), and the United States will continue to be most responsible for the human-made CO_2 increase for the next few decades, even though China's ongoing emissions will exceed those of the United States. Although a portion of human-made CO_2 emissions is taken up by the ocean, there it exerts a 'back pressure' on the atmosphere, so that, in effect, a substantial fraction of past emissions remains in the air for many centuries until it is incorporated into ocean sediments.

Hansen said that one Iowa coal plant, with emissions of 5.9 million tons of CO_2 per year and 297 million over 50 years, could contribute to tipping points in life systems and human behavior.

1 James E. Hansen, testimony before the Iowa Utilities Board, submitted Oct. 22, 2007.

"The biologist E.O. Wilson in 2006 explained that the 21st century is a 'bottleneck' for species, because of extreme stresses they will experience, most of all because of climate change. He foresees a brighter future beyond the fossil fuel era, beyond the human population peak that will occur if developing countries follow the path of developed countries and China to lower fertility rates. Air and water can be clean and we can learn to live with other species of creation in a sustainable way, using renewable energy. The question is: how many species will survive the pressures of the 21st century bottleneck?" Interdependencies among species, some less mobile than others, can lead to collapse of ecosystems and rapid nonlinear loss of species, if climate change continues to increase.

Hansen commented on the point that the natural world has undergone huge climate variations in the past.

"That is true," he testified, "but those climate variations produced a different planet." He went on to cite evidence about the melting of polar ice, the effects of changing sea levels and changes in salinity, and the climate changes that would make the extermination of numerous species inevitable.

Hansen said it is still possible to avoid dangerous climate change — but just barely. In his view, it would require phasing out coal use except at power plants that capture and sequester CO_2 and a moratorium on new coal-fired power plants — hard to imagine. It is the United States and Europe that have created the climate problem; unless we take dramatic action, he observed, "we have no basis for a successful discussion with China, India, and other developing countries."

Hansen is not easy to dismiss as the stereotypical, tree-hugging environmentalist. He is a trained climate scientist and a self-described moderate Republican.

The U.S. Climate Change Science Program report also stated that more volatile weather and declining sea levels will also have impacts on fuel deliveries and existing power plants in the U.S., which are heavily concentrated near the Atlantic Ocean.

And then there is the problem of all this violent weather battering off-shore oil rigs, which have to penetrate deeper and deeper to find more elusive oil. The DOE report says that off-shore production is at risk with increasing damage to platforms and pipelines. The report goes on to state that most oil producers prefer to repair the damaged oil rigs than to rebuild them to stand up to the increasingly turbulent weather expected to increase with global warming.

The program report also puts little faith in renewable energy sources which is also reflected in the Energy Information Administration energy report — and with good reason. Again, the effects of global warming are already at work because "renewable energy systems are vulnerable to damage by extreme weather events."

"Renewable energy sources are therefore connected with climate change in very complex ways: their use can affect the magnitude of climate change, while the magnitude of climate change can affect their prospects for use," according to the report.

Approximately 75 percent of electricity from renewable resources comes from hydro plants. Hydroelectricity projects also have their share of problems — again, these are related to water use. A reduction of the snow pack in the Western United States will mean a reduction in hydroelectricity. Most of the prime sites for hydroelectricity have already been dammed and cause ecological damage of their own by injecting warmer water into rivers and interrupting fish migration patterns. Rivers are running dry and many never make it to the sea because of industrial, energy, and irrigation usage. The water loss from some reservoirs in the United States and around the world is expected to increase from already alarming levels as the temperature rises. As the report succinctly puts it: there will be less water for all uses — water needed to irrigate crops that make the desert bloom in California, urban use in cities like Los Angeles, Las Vegas, and Tucson, and myriad other competing uses. And the politics of the West has always been the politics of water — with competition expected to become much more intense by 2030 both nationally and globally. The country's strategy to rely on its substantial coal reserves will also stress already stressed water resources around the country and around the world. This is called the "energy-water nexus."

Ethanol, once touted as a savior for the country's energy woes, is facing more and more obstacles as a viable replacement for fossil fuels. Most scientists from the outset have recognized ethanol for the boondoggle that it is. It takes more energy to produce ethanol than it produces itself. Corn is less effective than sugarcane, which is grown primarily in tropical climates. Converting land to ethanol production has led to less production of crops like soybeans. To address this opportunity in the global soybean market, more forests in countries like Brazil will be cut down to profit from soybean demand. Ethanol production, then, will contribute to global warming by robbing the world of woodlands that absorb carbon dioxide. The DOE report is also skeptical that ethanol will contribute significantly to the energy mix: "Processing the entire projected 2015 corn crop to ethanol (highly unrealistic, of course) would only yield about 35 billion gallons of ethanol, less than 14 percent of the gasoline energy demand projected for that year."

However, the DOE report does express concern about the domestic oil industry's ability to receive imported oil. More deepwater ports will be needed to accommodate supertankers and barges will have to navigate shallower rivers which will need to be dredged.

Extracting the last drops of oil from Alaska will also be challenging. The rise in Alaska's temperature is expected to be double the global average. Higher ambient temperatures make pipelines less efficient.

So where does that leave energy policy in the United States?

Even though polls in recent years have shown that the American public is increasingly alarmed about the effects of climate change, cooler heads prevailed at the DOE who, during the Bush Administration, were certainly not ready to press the panic button. This casual view from the government agency charged with overseeing the energy policies of the largest and most polluting economy, by far, on the planet saw no real need to rush into this climate change frenzy.

The report found climate change "cautionary rather than alarming" and that, fortunately, real problems are "some decades in the future" and would give the DOE "time to consider strategies for adaptation" to these irksome little problems. Because vital research is lacking, it's ... well ... time to do some research. Never mind that most of these environmental issues have been on the table for the past 50 years or longer. Climate change has been a huge concern for at least a decade. What *have* these people been doing? The major concern seemed to be that the energy sector will continue to survive in much the same manner as it adapts to the challenges of climate change for which it is primarily responsible so it can exacerbate the problem further.

Perhaps some of the cluelessness can be blamed on disingenuous comments made by officials such as former Secretary of Energy Samuel Bodman who said, on February 7, 2007, in response to the fourth Intergovernmental Panel on Climate Change report on global warming: "Even if we were successful in accomplishing some kind of debate and discussion about what caps might be here in the United States, *we are a small contributor* [emphasis added] when you look at the rest of the world."[1] In 2004, the United States contributed 22.2 percent of manmade CO_2 emissions, the most of any country; as of 2007, China's emissions were approximately equal.

While the United States has often been cited both within and without the country for its addiction to oil, making an analogy to addictive behavior when it comes to the energy sector isn't much of a stretch. The denial is evident. "I'm not so bad. At least I'm not that Chinese guy — he's *really* got a problem. Sure, I do more drugs and have for a long time, but he's been using just as much lately."

Will humans cease to exist by 2030? Probably not. That, however, could come a century or two later. Most reputable scientists believe that 450 parts per million CO_2 is a generally agreed upon threshold for the amount that would trigger dire atmospheric changes from which it would be difficult, if not impossible, to recover. The Earth's atmosphere is currently at 387 ppm (some estimates are slightly higher), a wildly unprecedented amount which is occurring at a wildly unprecedented speed. About 3 ppm is expected to be added yearly under the "business as usual" scenario. The atmosphere would reach this threshold by 2030.

---

1 Lisa Hymas, *"Bush's Climate Change,"* Grist Magazine, Feb. 12, 2007.

William Calvin, a professor at the University of Washington and author of the book *Global Fever: How to Treat Climate Change,"* said that the Energy Information Administration tracks closely with the IPCC "business as usual" scenario.[1]

"The most important consequences of a 'business as usual' path to 2030 are that we will miss the possibility of avoiding a 3 degree Celsius rise later in the century," Professor Calvin observes. "We are already at 0.8 degree Celsius fever and delayed effects alone will take us to 1.5 degrees Celsius even if we were to stop all emissions worldwide today. The present fever is already proving to be dangerous territory, given the widening of the tropics, large increases to land area in drought, wildfires, and major floods."

Spencer Weart is the director of the Center for History of Physics at the American Institute of Physics in College Park, Md. He is a historian specializing in climate science. Weart said the IPCC does not try to actually predict what will happen to climate, but rather gives a range for what might happen under different scenarios.

> These run from 'business as usual' with vigorous growth, to strong controls on emissions and population growth. The DOE projections of energy increase are similar to the 'business as usual' scenario of the IPCC, implying a strong rise in emissions of greenhouse gases. If it continues in this way for some decades, by the end of the century we would certainly see serious climate changes, and quite possibly a radical and catastrophic reorganization of the world's climate and ecosystems.

> However, little of this will happen by 2030. Because of our emissions over the past century, which have accumulated in the atmosphere and will remain there for centuries, we have already locked in a modest rise of global temperature over the next twenty years — even if human civilization were to disappear from the planet tomorrow. Whether our emissions are large or small, by 2030 there will be worse droughts, more intense storms (not necessarily hurricanes, but temperate-zone storms) and floods, the spread of tropical diseases and insect pests, etc. In the U.S. there will actually be benefits for agriculture over the next 20 years, with problems emerging only later. So the main reason for asking for restrictions now on the growth of energy use is because of its cumulative effect by the end of the century and indeed beyond. The 22nd century may seem very far off, but many people now living will be there to see it, and of course our grandchildren.

> I should also add that many experts doubt the DOE's projection is possible, since there are strong indications that the production of oil will level off, or perhaps has even already reached a plateau, and cannot keep increasing as they suppose.

Professor Lee Kump of Penn. State and an expert on coral reefs, said: "There are good indications that many ecosystems are already under stress, and the

---

1 William H. Calvin, *Global Fever: How to Treat Climate Change* (Chicago: The University of Chicago Press, 2008).

trend toward increasing CO_2 will only exacerbate the situation. I'm most concerned about coral reefs and other ecosystems based on organisms that produce their skeletons from calcium carbonate (limestone), because these ecosystems will be directly affected by the dropping pH (increasing acidity) of the ocean as CO_2 increases. But there is no one CO_2 concentration threshold for these or any other ecosystem; each has its own sensitivity to CO_2 build-up and attendant climate change."

The point is not that the Earth would then be unfit for human habitation. This will mean that the damage has been done, cannot be undone, and the system becomes dysfunctional. The lag time will be considerable. Today's young adult will experience a substantially diminished environment and a degraded quality of life. *Their* children will face a world that is much, much worse where survival will become an issue. If the 450 parts per million marked is breached, the window is closed, the horse is out of the barn, the ship has sailed, and the canary in the coal mine is dead. The time will have come to find a more hospitable planet.

However, that is not to say that the world of 2030 will only present environmental challenges. The next generation will face economic, political, and social turmoil, mass migration, and cultural conflict. It will also become clear that American-style capitalism is on a crash course with the Earth's carrying capacity. The double-whammy of environmental degradation and unsustainable population growth will conspire to overtax essential resources.

The sad fact is that there are no technologies in the pipeline even remotely capable of replacing fossil fuels. The world's two largest economies — the United States and China — are both coal rich and oil poor. The availability of coal coupled with higher oil prices will make coal too attractive to ignore. All of the small steps undertaken by environmentally-conscious individuals are overwhelmed by the increase in the energy sector. No substitute for oil has been found to fuel the huge transportation sector. Few of us are ready to forego a car or those airplane trips to visit our scattered families. Few of us will accept a standard of living lower than our parents. The middle class will struggle to keep up, to beat back the "fear of falling." These concerns are hard-wired into the human psyche even during times of economic recession and financial hardship. For all but the fortunate few, status will increasingly be maintained with borrowed funds until the inevitable day of reckoning and the recent economic troubles have shown how close we are to that day. The United States is heading for social stratification that will have more in common with Bolivia than post-World War II America.

The world faces a no-win situation.

James Lovelock, an independent scientist in the United Kingdom and a fellow of the Royal Society, told *The Independent* in December 2005:

> The climate centers around the world, which are equivalent of a pathology lab of a hospital, have reported the Earth's physical condition and the climate specialists see it as seriously ill, and soon to pass into a

morbid fever that may last as long as 100,000 years. I have to tell you, as members of the Earth's family and an intimate part of it, that you and especially civilization are in grave danger.[1]

...We (in the United Kingdom) could grow enough to feed ourselves on the diet of the Second World War, but the notion that there is land to spare to grow biofuels, or be the site of wind farms, is ludicrous. We will do our best to survive, but sadly I cannot see the United States or the emerging economies of China and India cutting back in time, and they are the main source of emissions. The worst will happen and survivors will have to adapt to a hell of a climate.

Dr. Lovelock gave another one of his rare interviews to *Rolling Stone* magazine in October, 2007.[2]

Our future is like that of the passengers on a small pleasure boat sailing quietly above Niagara Falls, not knowing that the engines are about to fail.

Although he said that he is not an alarmist by nature, Dr. Lovelock also observed, "You could quite seriously look at climate change as a response of the system intended to get rid of an irritating species: us humans. Or at least cut them back to size."

Because thinkers from various disciplines can see the problems the planet faces now and in the future, a variety of essays were published under the ominous title of *Global Survival* and viewed from various perspectives. Some of the comments are as follows:

According to John H. Herz:[3]

There is the growing impact on public opinion by economically and/ or ideologically focused groups interested in the maintenance of a favorable status quo that, while endangering the future of the world, promote their narrow interests. Businesses interested in a deregulated economy free from government interference and environmental regulations, and the military interested in job security guaranteed by continually increasing armament programs, are examples of this.

In our era of 'public relations,' vested interests utilize ever more sophisticated means of influencing, if not controlling, public opinion and attitudes. This partly explains the failure of movements and organizations concerned about threats to global survival to impart this awareness to larger publics or to mobilize for more than short periods of time. Policy is occupied more with creating comforting public images than with accomplishing results such as resolving problems through mutual compromise.

1 Ian Irvine, *"James Lovelock: The Green Man,"* The Independent [U.K.], Dec. 3, 2005.

2 Jeff Goodell, *"The Prophet of Climate Change: James Lovelock,"* Rolling Stone, October 2007.

3 John H. Herz, essay *"On Human Survival: Reflections on Survival Research and Survival Policies,"* Global Survival, editors Ervin Laszlo and Peter Seidel (New York: SelectBooks, Inc., 2006) pp. 12, 13, 14, 18, 19, 22 and 23.

"In contrast to the nuclear threat, the great complexity inherent in the various facets of the ecological threat makes it less easy to arrive at Survival Research-type conclusions. As I have pointed out, specific developments such as deforestation and desertification in sub-Saharan Africa or overpopulation in Third World metropolitan areas may not appear as dangers to the global ecosystem, per se. Only in their totality will they turn out to be life-threatening for all of us. Thus, the industrialization of Third World countries in conjunction with the absence of effective population controls has meant resource depletion, urbanization without raising the living standards of the majority of the population, and migration of rural populations to megacities. This causes the destabilization of regimes that try to defend themselves against overthrow through import of arms (at the expense of resources for genuine development) and incessant, endemic internal or external, minor or major warfare, often accompanied by the laying of landmines that make entire regions uninhabitable. The overall world arms production and trade, and the ensuing armaments and arms races, not only raise the specter of nuclear war through escalation but, with their tremendously accelerating waste of non-renewable resources, pose a direct threat to ecological survival. This affects not only Third World countries, but the industrialized countries as well. The ever-growing portion of GNP deflected to military purposes results in a vast waste of human and material resources, neglect of infrastructure, erosion of the welfare state and ensuring economic and social polarization, and so on."

"There has been a world-wide change of ideology and policies from a search of domestic and international regulation and planning of 'better worlds' (that since the age of industrialization created welfare states and measures of environmental protection and preservation) to the ideology and policies of the regulation-free market system, with non-interference of state and international organizations in the affairs of globalized economies. Laissez-faire means non-pursuance or even abrogation of efforts and measures toward environmental protection. In the United States, the Bush regime has been abrogating whatever modest measures have been taken by preceding administrations, whether in regard to diminishing the emission of carbon dioxide gases or the destruction of forests or other land in the interest of free exploitation by timber or oil-drilling interests, not to mention reducing the already insufficient aid to Third World nations to create even minimally sufficient infrastructures for sustainable human habitat."

"Generally speaking, the laissez-faire ideology in an age of rampant, unyielding globalization implies a tendency on the part of vested interests to spread this ideology and the corresponding attitudes to the general public through control of the media and, increasingly, even the educational institutions forming the minds of the new generations. This even extends to influencing academia."

"Radical turns have happened in human history before — now it must happen globally. If the species is not able collectively to act so as to provide for its continued existence, then indeed its final fate will be indicated by the words *Exeunt omnes — Finis.*"

According to David and Marcia Pimentel:[1]

The natural ecosystem and the diverse species it contains serve as a vital reservoir of genetic material for future development in agriculture, forestry, pharmaceutical products, and biosphere services. Yet with each passing day an estimated 150 species are eliminated from the planet because, as human numbers continue to increase, their diverse activities expand into natural ecosystems. The activities include deforestation, soil and water pollution, pesticide use, urbanization, development and extension of transport systems, and industrialization. The rate of extinction of some species under these conditions is 1,000- to 10,000-times faster than occurs in natural systems. The fact that humans use more than 50 percent of the solar energy captured by the entire planet biomass to meet their food and fiber needs is a major cause of these high rates of extinction. This means that the photosynthetic biomass available to maintain vital natural biota is significantly reduced and biodiversity is greatly reduced.

Humans have no technologies which can substitute for the food, medicines, and diverse services that plant, animal, and microbe species provide. For example, one third of the human food supply relies either directly or indirectly on effective insect pollination. Each year, honey bees and wild bees are essential in pollinating about $40 billion worth of U.S. crops, in addition to natural plant species. Including pollination, the economic benefits of biodiversity in the United States are an estimated $300 billion per year, and nearly $3 trillion worldwide. Indeed, plants, animals, and microbes also carry out many other essential activities for humans, recycling manure and other organic wastes, degrading some chemical pollutants, as well as purifying water and soil.

Despite all projections about human population growth, no one really knows exactly how large the human population will be in 50 years. We know the 6.2 billion people already on Earth are stressing the Earth's land, water, and biological resources and polluting the environment. We know that more than 3 billion malnourished people are too many.

If the human population continues to increase and exhaust the Earth's natural resources, nature will control our numbers by disease, hunger, malnutrition, and violent conflicts over resources.

According to Jerome H. Barkow:[2]

Global warming is affecting us massively; we are destroying much of our planet's biodiversity.... But the evolved mechanisms invoked are those that deal efficiently only with issues of coalitional politics and reputation, the kinds of longer-term problems that our Pleistocene ancestors *would* have faced. Only when a vague collective worry becomes

---

1 David and Marcia Pimentel, ibid., pp. 41, 46.

2 Jerome H. Barkow, essay *"Biology is Destiny Only if We Ignore It,"* Global Survival, editors Ervin Laszlo and Peter Seidel (New York: SelectBooks, Inc., 2006), p. 73.

a short-term personal threat to ourselves or those close to us do we react with the force these problems merit. We become members of a concerned and activist minority, the ethnocentrism reaction kicks in, and we enjoy a sense of moral superiority because we are now an in-group engaging in collective action, and we lament the moral inferiority of those who do not realize that their house is on fire.

James E. Alcock adds:[1]

[D]ie we ultimately must. And to deal with the existential angst that this realization produces, societies everywhere have constructed belief systems — religions, cults, philosophies — to assuage anxiety about the prospect of personal demise....

And so it has been throughout human history, this overriding concern for personal survival. However, until recent times, we have never had to think very much about threats to the survival of our species, or about the sustainability of the physical environment upon which we depend for our existence. Yet, largely because of products of the astounding scientific and technological progress of the past century, and the major societal changes that they have helped produce, we now face a plethora of planetary perils: threats posed by overpopulation in some regions and precipitously falling birth rates in others, deleterious environmental changes, resource depletion, pollution, nuclear and biological weapons, and new forms of pestilence.

Given the ubiquitous human concern for personal survival, one might expect the citizens of the world to be just as concerned about their collective survival, and to demand that their leaders work together to confront these perils. Such is not the case. Concerns about collective survival fall for the most part on deaf ears. Why should this be so? No doubt a number of factors account for the disconnection between concerns about personal and collective survival, but a key element is *belief*.

Some beliefs also serve the important function of reducing anxiety. The worry that follows a serious injury is reduced, at least for a time, if we believe that modern medicine will cure our ills.... The fears raised by warnings about the dangers of greenhouse gases, global warming, and acid rain are reduced if we believe that these warnings are gross exaggerations, or that, if the problems are real, scientists will solve them for us. Such mitigating beliefs are based partly in hope or blind faith and partly in denial.... A belief based in denial may also reduce anxiety in the short term, but it is more likely to be maladaptive in the long run, because instead of motivating action to overcome the threat, it calms people into doing nothing.

According to Richard B. Norgaard and Paul Baer:[2]

---

1 James E. Alcock, essay *"Belief and Survival,"* Global Survival, editors Ervin Laszlo and Peter Seidel (New York: SelectBooks, Inc., 2006), pp. 86, 87, 88, 94, 95 and 97.

2 Richard B. Norgaard and Paul Baer, essay *"Seeing the Whole Picture,"* Global Survival, editors Ervin Laszlo and Peter Seidel (New York: SelectBooks, Inc., 2006), pp. 141, 142, 145 and 156.

The controversy, in part instigated and sustained by corporations with an interest in maintaining a fossil fuel-based economy, has confused the public and politicians in the United States. Many now feel that it is only rational to delay action until the science is sufficiently strong. Thus we have a problem of how we understand and how we translate understanding into action that threatens human survival as we know it.

The politics of global survival is largely a game of individual and corporate greed played on a field of human short-sightedness. The particular political maneuvers in the debate over climate change, however, also clearly play off historical understandings of how science is supposed to work and to connect to the policy process. To some extent, powerful interests are misconstruing appropriate principles about how science works. At the same time, scientists and philosophers have not adequately addressed how we understand across disciplines.... Seeing environmental problems as systemic problems between social and environmental systems, especially when the problems are also global, forces us to rethink how we know.... Establishing deeper and richer understandings of how science works would reduce the opportunities for corporate greed to trump the understanding of scientists.

[S]cientists within disciplines have become increasingly isolated and have not updated — through keeping up with developments in other disciplines — how their separate knowledge fits into the whole picture. This raises serious questions about how the unity of science actually unifies across multiple minds. And even if it could be unified within science, one has to ask how the knowledge inherent in the unity can be drawn upon by the public at large.

Human survival will depend on scientists and the public at large seeing the whole picture. We probably have most of the pieces of the whole picture already in the form of disciplinary knowledge, or the pieces could be seen within the disciplines once the connections between the disciplines are made appropriately. Thus, we are arguing that building understanding of the interconnections across the disciplines is a critical element of seeing the whole picture.

Kenneth E.F. Watt says:[1]

[While] it may seem like blatant nonsense to suggest that it is even worthwhile to consider the survival of civilization, [i]n fact, there are many arguments based on fact, logic, reasoning by analogy from other civilizations, or studying the dynamic behavior of systems to make one very alarmed.

Then he cited Chris Matthews and Tim Russert talking about the 'Happy Face' phenomenon.

Matthews stated that amongst a set of politicians competing for any office, a safe prediction is that the winner will be the one who most

---

1 Kenneth E.F. Watt, essay *"What Can the Systems Community Contribute to Ensure the Survival of Civilization?"* Global Survival, editors Ervin Laszlo and Peter Seidel (New York: Select-Books, Inc., 2006), pp. 159, 160, 161, 162 and 169.

consistently presents a happy face, and an optimistic vision of the future characterized by growth of everything, free of any threats. The American electorate severely punishes politicians who attempt to present any unpleasant information....

Highly sophisticated and meticulously planned psychological manipulation of the electorate is now a key feature of a winning strategy in elections. The United States population has now become filled with a vague dread and terror which affects behavior. This terror has been very carefully communicated to them, in association with the idea that one political party will be more capable of protecting the population than the other.

He also illustrated a revealing shift in the society we live in.

The number of people awarded all Bachelor's, Master's and Doctor's degrees in the United States grew by 61 percent from 1970 to 1999; the number incarcerated in all jails, prisons and penitentiaries grew by a factor of six. All kinds of other social statistics reflect this competition between two major industries, one of which is the clear winner. For example, there is enormous growth in the number of attorneys, judges, police, guards, wardens, detectives, and sheriffs in contradistinction to much slower growth in numbers of professors, doctors, and teachers. The number of Bachelor's degrees awarded in education in 1999 was 60 percent what it was in 1971. A similar picture emerges when one explores the trends in degrees awarded in fact-based fields as opposed to belief-based fields.

Andy Bahn and John Gowdy stress the central role of the market economy in shaping our the ecological and social conditions:[1]

"Given on planet Earth, it is natural that the field of economics should be a leading player in Survival Research. The growth of the market economy, including increased consumption rates, rapid urbanization, and growing income disparities within and between nations, lies behind all the negative environmental trends of the past 30 years. All these important trends, then, fall within the domain of economics and should be the focus of intensive efforts by economists to understand the human impact on the natural world.... Contemporary economic theory is both a description of how stylized markets allocate resources, and an ideological justification for the superiority of the unfettered market economy. Survivability research requires both an understanding of how market economies operate, and an understanding of how the force of market exchange might be diverted from its current destructive path to one of environmental and social sustainability. The new field of ecological economics explicitly addresses the survivability question.

---

1 Andy Bahn and John Gowdy, essay *"Economics Weak and Strong: Ecological Economics and Human Survival,"* Global Survival, editors Ervin Laszlo and Peter Seidel (New York: SelectBooks, Inc., 2006), pp. 176, 180, 183, 184 and 186.

...But in the past two hundred years the market economy has evolved into something very different than any past human culture experienced. Understanding how markets work and how the market economy has changed the relationship between humans and the natural world is essential to the survival of our species....[D]decisions now are characterized by a high degree of uncertainty, limited information, and a very short time frame in which to operate.

In view of the importance of climate change and biodiversity loss to our very existence as a species, it seems incredible that we cannot take the necessary steps to address these problems. The fact that world leaders hold the environment in such low priority shows the power of the ideology of the global market economy — an ideology glorifying consumption, production, and economic wealth over long-term survivability.

J.R. McNeill[1] borrows from Churchill's aphorism about democracy, which he said was the worst of all political systems, except for all the others.

> History, one might say, is the worst of all possible guides to the future — except for all the others.

> The first thing to understand is the bizarre, anomalous, and thoroughly unsustainable character of the recent past, especially the last half century. ...[T]he dizzying pace of growth since 1950, came at a price; or perhaps better put, at two prices. The first is that the growth was (and remains) sharply unequal, so that the modern age was (and remains) one of differentiation, of growing divides between rich and poor. This of course may be seen as a good thing because it means that some people, indeed more than ever before, live free from poverty. But if one prefers equality, it is decidedly a bad thing. At any rate, it contains the seeds of further instability because in the present age of cheap information, the world's poor will not remain ignorant of their plight, and many of them will not remain resigned to it. They will respond with ambition, hard work, migration, revolution, crime, and other forms of initiative, all intended to reconcile the yawning gap between their circumstances and their wishes. Growing economic inequality in an age of cheap information is a combustible recipe.

Indeed,

> Driven by energy use, population growth, urbanization and technological change, among other things, we have lately created a regime of perpetual disturbance in global ecology. In such unsustainable circumstances we have prospered mightily as a species, growing in number, in wealth, and in the share of the biosphere's energy and materials that we can turn to our own uses. It is as if we had craftily shaken things up so as to help ourselves at the expense of other species. But ... "The dramatic changes in the human position within the Earth's ecosystems ... continue to add to the ecological tumult: global ecology changes as rapidly

---

1 J.R. McNeill, essay *"Historical Perspectives on Global Ecology,"* Global Survival, editors Ervin Laszlo and Peter Seidel (New York: SelectBooks, Inc., 2006), pp. 191, 193, 196, 198, 201, 202 and 203.

as it does in part because ideas and politics, from an ecological point of view, change so slowly.

And he notes,

> Communism shared one of its strongest ideological commitments with capitalism: the growth fetish. This was a flexible and seductive creed, appealing to almost everyone in authority, because economic growth hid a multitude of political sins. Populations would put up with corruption, vast inequalities, and repressive police states if they were confident that in the years to come they would be materially better off than at the moment.

> While communism is at least temporarily dead as a big idea, the traditional emphases of nationalism and the growth fetish remain with us in the 21st century, and will continue to help fashion new dramas of ecological change.

According to Richard D. Lamm:[1]

> Most of human experience is on the side of both continued population and economic growth culture. The world of growth has succeeded brilliantly. It allowed survival in a harsh world. It has brought health, wealth, increased life expectancy, leisure, and most important — freedom. Growth has approached the status of a religion. Sociologist Peter Berger points out: 'Development is not just a goal of rational action in the economic, political, and social spheres. It is also, and very deeply, the focus of redemptive hopes and expectations. In an important sense, development is a religious category. Even for those living on the most precarious margins of existence, development is not just a matter of improved material conditions; it is at least also a vision of redemptive transformation. But even in our religious fervor, we must ask, 'can it last?' Is this a sustainable vision? Is growth the permanent secret to success for societies?

> Power tends to corrupt all human institutions, but the U.S. Constitution and its balance of powers have worked well to deal with corruption of abuse of power. The two-party system plays a real role in this process in debating issues and exposing self-interest. To a remarkable degree, we have enjoyed a self-correcting system. Now, however, the special interests have found a way to avoid this self-correcting system. They have taken over both political parties. They advance their coercive agenda by electing all or most of those who make the rules. It does not matter whether someone is a Republican or a Democrat — do they support and defend your particular interests? Special interests can often pass and stop legislation at will. Their actions are becoming increasingly blatant.... We use the resources of the uncomplaining many to satisfy the complaining and self-interested few. Eventually, this is a political Ponzi scheme that is bound to crash.

> The first step in solving a problem must be to correctly identify the nature of the problem. We have an institutional problem more than a

---

1 Lamm, ibid., pp. 208, 214, 217 and 219.

political problem. Both political parties are for sale; both are hopelessly compromised by special interest money. Our political institutions, instead of being part of the solution, have become part of the problem. Only a new party, unencumbered by the past, can take the money out of politics or reduce its caustic influence.

A number of new questions of public policy are being raised. What is a country's demographic destiny? Can the world sustain the standard of living that all countries are striving for? Can we have infinite growth in a finite environment? Is humankind losing its margin of error and setting itself up for some calamity? We never raised the issue before because we didn't recognize it as an issue.

Some new issues become new issues because we recognize, suddenly, that they are amenable to change. We never thought that health was subject to human control. We either got well or died depending on 'God's will.' Then we found medicine. We always thought the number of children a woman had was a matter of 'God's will,' but then we learned about contraception and recognized that this too was subject to human control. What seemed a fixed fact was, in fact, a policy choice — "what is our demographic destiny?"

The current dilemma and challenge is not only that we are faced with unprecedented change; additionally, yesterday's solutions have become today's problems. The transition from growth to sustainability will be one of the greatest transitions of history. Our economic, religious, social, and political systems all are built on the growth model. It may be beyond the capacity of our institutions to handle this transition without massive disruption. But no generation gets to pick the challenges it is faced with. Sustainability is the current challenge, and it will not go away.

According to Christopher Williams:[1]

Management theories of leadership are practical, but apply essentially to commercial interests, which are often antithetical to the interests of humanity and to our survival.

Like it or not, contemporary political leaders are going to win more votes through preserving the American way of life than through preserving humanity. And while our only form of world governance is international rather than global, the formal views of the international leaders are likely (and not wrongly) to reflect those of their paymasters.

Any national education system is to some extent a reflection of the ideology of those in power.... The Taliban close private girls' schools, rule that the other schools could not instruct girls older than eight, and they could only teach the Koran. In Russia, the supreme mufti Ravil Gaynutdin has expressed concern about Islamic colleges that are promulgating the teachings of Wahabism. This 18th century movement calls on its followers to treat all other Muslims as enemies. Current textbooks teach that anyone who is not a Wahabi should be killed. That ideology

1   Christopher Williams, essay *"Educating World Leaders,"* Global Survival, editors Ervin Laszlo and Peter Seidel (New York: SelectBooks, Inc., 2006), pp. 239, 244 and 248.

amounts to proposing the extermination of most of humanity. Islamic fundamentalism is not the only source of abuse. In some southern states of the U.S., creationists impede knowledge about evolution and Charles Darwin."

The threat to survival arising from the ideology of unbridled progress is latent and insidious, and so far has attracted little comment.

According to James Lovelock:[1]

Take a good look around in any bookstore or library. The books there may be well written, entertaining, or informative, but most of them deal with superficial and rather evanescent topics. They take a great deal for granted — the hard-won scientific knowledge that gave us the safe and comfortable lives we enjoy today. Most of us are so ignorant of the facts upon which science and our scientific culture are founded that we give equal place on our bookshelves to nonsense like astrology, creationism, and junk science. At first, such things served just to entertain us, or indulge our curiosity — we didn't take them seriously. Now they are often accepted as fact, and given "equal time."

Without drastic changes in policy — both domestically and globally — "Business As Usual" signs will be replaced by "Closed Until Further Notice."

1 James Lovelock with Peter Seidel, essay "A Primer of Civilization," Global Survival, ed. Ervin Laszlo and Peter Seidel (New York: SelectBooks, Inc., 2006), p. 259.

CHAPTER SIX — DENIERS

"All of our exalted technological progress, civilization for that matter, is comparable to an axe in the hand of a pathological criminal."
— Albert Einstein in a letter to friend Heinrich Zangger

"The scientific debate is closing [against us] but not yet closed. There is still a window of opportunity to challenge the science."
— Republican pollster Frank Luntz

"I'm optimistic. Our response to global warming needs to be sweeping, but it needn't be traumatic. Instead, we'll create new industries and open new markets."
— Senator Ron Wyden, D–OR, April 2007

There is no other way to put it. Corporations, and corporately-financed think tanks and lobbyists, have been remarkably successful in convincing a significant number of Americans to act against their own interests. Usually under the specious banner of "freedom," these corporate interests, and lackeys funded by corporate interests, have successfully created a well-funded machine to assure no significant action to combat climate change will be undertaken now or in the decades to come.

To suggest that American-style capitalism or globalization may have any finite limits in an environment severely under stress has been painted as akin to communist leanings and economic, social, and religious apostasy. The strange alliance between corporate greed and religious fundamentalism has melded to create a political climate dangerous to this and future generations. Scientists may be good at science, but they are terrible in the political arena and are dependent upon the very resource that corporations possess in abundance — money.

The strategy was more amazing as the economy grew, profits were up, and the average American saw none of the profits, endured job losses, and was abetting the dismantling of the country's middle class. The term "useful idiots" comes to mind. Both parties — Republican and Democratic — have become too reliant on corporate funding with short-term profits funding short-term political elections. U.S. politicians have come to realize that re-election is impossible if the candidate proposes to either raise taxes or cut programs — an impossible proposition in the long run. The economic pain will be borne by subsequent generations.

Any serious discussion about climate change is perceived as a threat to both the corporate class and the now-mistrusted and discarded George W. Bush Administration. The heady, triumphant days of the early Bush years are now relegated to the trash bin of history. Government became passé and unregulated free markets were the wave of the future. Iraq was supposed to be a shining example of this wisdom. Climate concerns threaten to change all that. The economic stakes are enormous and the policy of unlimited growth — the very foundation of American capitalism — is beginning to draw more fire. A world economy that surrendered to this American economic umbrella is now suffering for economic woes "made in America." Because unlimited free-market economics has reached the status of a secular religion amongst the corporate and political elite, the war against science is worth fighting. The decision to combat the *idea* of climate change has been much more successful and cheaper than actually addressing climate change. The American public has neither the time nor the energy to sort out fact from fiction. As long as global warming can maintain the status of a "debate," no serious action is necessary. Media interviews are reduced to "he said/she said" affairs. The climate change deniers are using the same stall tactics that proved so effective for tobacco companies — stall as long as possible.

A bevy of these right-wing think tanks with noble-sounding names (the latest being the Consumer Energy Alliance, a consortium of corporations) have been subsidized by the energy industry, especially ExxonMobil and the coal industry. Funding sources can only be discovered after some investigation, pulling away the curtain in the corporate Land of Oz. The net result is that these vested interests will almost certainly delay critically needed action until the opportunity to address the problem has passed.

The issue is easily resolved: simply employ the old axiom of "follow the money." Who will gain and who will lose?

According to Sharon Begley, a writer for *Newsweek*, this well-funded and well-coordinated effort by contrarian scientists, free-market think tanks, and industry has been under way since the late 1980s.[1] The objective has been to sow the seeds of doubt about climate change. At first, greenhouse gas deniers argued that the world was not warming. This strategy had to be dumped once the overwhelming

---

1 Sharon Begley, "*Global Warming Deniers: A Well-Funded Machine*," Newsweek, Aug. 13, 2007.

majority of world's scientists proved otherwise. Next came the arguments that the measurements were flawed. This turned out to be true, but not in the way these groups might have hoped — climate change was happening more rapidly than even scientists or computer models predicted. The next tactic was to claim that the warming was natural, not a product of man-made emissions. This was also debunked. The current argument is that the effects of climate change will be a minor event, relatively harmless and even, in some cases, beneficial.

These ideas, no matter how absurd and devoid of any scientific inquiry, have been remarkably easy to sell. Not acknowledging a problem is infinitely easier than addressing it. Excess consumption, they argue, is not a problem at all. No pain is required; no excessive habits need to be changed.

In 2006, polls found that 64 percent of Americans thought there was "a lot" of scientific disagreement on climate change; only one-third thought global warming was "mainly caused by things people do." Majorities in Europe and Japan said that humans were altering the world's climate. A 2007 *Newsweek* poll found that 42 percent of Americans said there is a lot of disagreement that human activities are a major cause of global warming. Only 46 percent said the greenhouse effect is being felt today.

"As soon as the scientific community began to come together on the science of climate change, the pushback began," said historian Naomi Oreskes of the University of California, San Diego. Individual companies and industry associations — representing petroleum, steel, autos and utilities, for instance — formed lobbying groups with names like the Global Climate Coalition and the Information Council on the Environment. ICE's strategy called for greenhouse doubters to "reposition global warming as a theory rather than a fact." ICE and the Global Climate Coalition lobbied hard against any international treaty on climate change and were joined by another player: the George C. Marshall Institute, a conservative think tank. The gist of their argument against man-made global warming was that the sun was putting out more energy, even thought this explanation fell far short of explaining the extent of the changes documented by scientists.

Think tanks were always on the lookout for contrarian scientists who could be persuaded — financially or otherwise — to join their camp. Like the doctors over the years who have been persuaded to shill for pharmaceutical companies, one think tank offered scientists a $10,000 prize to write articles debunking a report from the IPCC.

Growing international concern during the Clinton Administration only served to ramp up the campaign. Under the George H. W. Bush Administration, the energy industry was able to avoid mandatory cuts — regulations that were anathema to the tenets of strict free-market champions. However, even nemesis Al Gore was relatively silent about pressing environmental issues during his run for the presidency. The denial machine had been so effective that Gore knew a tough stance against climate change problems was a political loser because of the fog enveloping the issue emanating from conservative think tanks. Gore would have to wait until he was out of office to seriously address the issue.

Clinton, attracted as he was to the high-flying movers and shakers at Davos, proved to be no champion of the environment and did not even try to persuade the Senate to ratify the Kyoto treaty. Clinton was a prime example of how ineffective Democrats are when politicians are faced with the prospect of bucking corporate interests, lobbyists, and money. Instead of addressing climate change in any serious fashion, Clinton did champion NAFTA which allowed corporations to flee to Mexico to avoid environmental regulations.

Patrick Michaels, a climatologist at the University of Virginia, wrote extensively on climate change and called "apocalyptic environmentalism" the "most popular new religion to come along since Marxism." He admitted in 1995 that he had received more than $165,000 from corporate interests. The coal industry's Western Fuels Association paid Michaels to produce a newsletter called *World Climate Report*, which regularly trashed mainstream climate science.

The denial machine was very good at painting anybody expressing concern about climate change as dupes of communist ideology, enemies of the American way of life, fuzzy-headed intellectuals, and not real men with the proper "can-do" attitude but, rather, pessimists and purveyors of gloom and doom. The real Americans drove Ford F-150s or Hummers, drank manly amounts of beer, and had no problem with tossing the empties out the truck windows. And the tree huggers could go fuck themselves.

Meanwhile, the IPCC reports compiled by 2,500 scientists were becoming increasingly alarming. One key word used by the deniers was "international." They tapped into the segment of U.S. society that sees any global cooperation as synonymous with the vilified U.N., one-world government, and the stealthy black helicopters that were bound to follow. The IPCC concluded that "the balance of evidence suggests a discernible human influence on climate." This consensus was not good news for what had by now become the denial industry. The most damaging information was the effect of the "business-as-usual" scenario which, by any standards, was unsustainable. If the idea that substantial environmental damage would be caused (and had already been caused) to the Earth's ecosystem, industry would be obligated to curb emissions, retrofit existing plants, and adopt other safeguards that industries found onerous. These costs would be mind boggling. But, according to scientists, the alternative was worse, much worse. The stakes were high.

"There was an extraordinary campaign by the denial machine to find and hire scientists to sow dissent and make it appear that the research community was deeply divided," Dan Becker of the Sierra Club told *Newsweek*. These hired guns blitzed the media. Driven by notions of fairness and objectivity, the press "qualified every mention of human influence on climate change with 'some scientists believe,' where the reality is that the vast preponderance of scientific opinion accepts that human-caused emissions are contributing to warming," according to William Reilly, former head of the Environmental Protection Agency under President George H. W. Bush, who had failed to get more than voluntary cuts under that administration.

Right-wing talk-radio icon Rush Limbaugh played to his base by telling listeners as late as 2007 that "more carbon dioxide in the atmosphere is not likely to significantly contribute to the greenhouse effect. It's just part of the hoax." Conservative media support was effective. In a 2007 *Newsweek* poll, 42 percent said that the press "exaggerates the threat of climate change."[1]

In April 1998 representatives of the deniers, including the Marshall Institute and Exxon, met at the American Petroleum Institute's Washington headquarters. They proposed a $5 million campaign, according to a leaked memo, to convince the public that the science of global warming was riddled with controversy and uncertainty. The strategy was to train up to 20 "respected climate scientists" on media skills and public relations in order to raise "questions about and undercutting the 'prevailing scientific wisdom,' " especially attacking "the Kyoto treaty's scientific underpinnings" so that elected officials "will seek to prevent progress toward implementation." The plan was never implemented after it was leaked to the press.[2]

Under Clinton, the majority Republican Congress for six out of eight years was more than happy to collude with the denial machine. Joining GOP advocates were powerful Democrats like Rep. John Dingell of Michigan to protect the auto industry from mandatory CAFE (Corporate Average Fuel Economy) standards. Dingell, however, was an exception and global warming became a bitter bipartisan issue. Republicans received far more cash from industry than Democrats. As a result, no substantial debate on climate change ever reached the floor of Congress. Democratic Senator John Kerry called this active inaction "pure, raw pressure combined with false facts."

Enter the George W. Bush Administration in 2001 which initially was viewed by the deniers with a dose of skepticism. As a candidate, Bush had pledged to cap carbon dioxide emissions. The Competitive Enterprise Institute heard rumors that Bush was going to reiterate his campaign pledge after taking office. The think tank mustered allies to exert as much pressure as possible to have the pledge deleted from an upcoming speech. It was.

---

1 PollingReport.com: Environment, Newsweek Poll conducted by Princeton Survey Research Associates International, Aug. 1-2, 2007.
2 John H. Cushman Jr., *"Industrial Group Plans to Battle Climate Treaty,"* The New York Times, April 25, 1998.

Despite differing credentials in climate science and science in general, the public could not differentiate one scientist from another. *Newsweek*'s Sharon Begley wrote that "the public didn't notice. To most civilians, a scientist is a scientist."

"In the House, the leadership generally viewed it as impermissible to go along with anything that would even imply that climate change was genuine," David Goldston, who served as Republican chief of staff for the House of Representatives science committee until 2006, told *Newsweek*. "There was a belief on the part of many members that the science was fraudulent, even a Democratic fantasy. A lot of the information they got was from conservative think tanks and industry."

In the Senate, the deniers received a boost when Senator James Inhofe, a Republican from Oklahoma, took over as chairman of the environment committee. Inhofe was a denier's dream, announcing that "man-made global warming is the greatest hoax ever perpetrated on the American public." Inhofe, who consistently cites the Bible on various political issues, has compared environmentalists to Nazis and the EPA to the Gestapo. He also claimed that the Weather Channel was the sinister force behind the global warming hoax. Incidentally, the oil and gas industry is his number one political donor, giving him $972,973 since 1989. Inhofe held hearing after hearing, inviting prominent deniers to make their case. These included a scientist whose study was partly underwritten with $53,000 from the American Petroleum Institute.

"I was hearing the basic argument of the skeptics — a brilliant strategy to go after the science," said Tim Profeta, a former staff member for Senator Ted Stevens of Alaska who became a director of an environmental policy institute at Duke University. "And it was working."

The deniers were remarkably successfully in killing Congressional bills by keeping track of what federal scientists wrote and said. "If (the government) would have presented the science honestly, it would have brought public pressure for action," said Rick Piltz, who joined the federal Climate Science Program in 1995. By appointing former coal and oil lobbyists to key jobs overseeing climate policy, he found, the administration made sure that didn't happen. Officials made sure that every report and speech cast climate science as dubious as an excuse for inaction. Ex-oil lobbyist Philip Cooney, working for the White House Council on Environmental Quality, edited a 2002 report on climate science by inserting phrases such as "lack of understanding" and "considerable uncertainty." Another short section in another report was cut entirely. The White House "directed us to remove all mentions of it," says Piltz, who resigned in protest. Lobbyists praised Cooney for his creative editing.[1]

No one can say that the climate change deniers can't take a punch. As the scientific data on climate change mounted, they seized on a new theme: global

---

1 Andrew C. Revkin, "*Bush Aide Softened Greenhouse Gas Links to Global Warming*," The New York Times, June 8, 2005.

warming is really nothing to worry about. This outlook was still reflected in reports released by the White House-endorsed Climate Science Program until the end of the Bush Administration.

Over the years, information pounded into the heads of politicians by the energy industry remains embedded in their brains — or adding up in campaign coffers. ExxonMobil had cut back on the estimated $19 million it had given to the denial machine. This change of direction, many suspected, was because of the anticipated Democratic presidential victory of Barack Obama in 2008 and ExxonMobil wanted a place at the table when environmental policy was discussed.

As the *Newsweek* article points out, major discussions will have to center around how much money Americans are willing to pay to combat the worst of climate change. While polls show that an increasing number of Americans are very concerned about the effects of climate change, few would be willing to spend money on the problem. This was true even before the hefty price tag for recent corporate bailouts. The *Newsweek* poll, for instance, found less than half in favor of requiring high-mileage cars or energy-efficient appliances and buildings. Americans will likely begin to address the problem when the increasingly violent and erratic weather becomes even more visible — which, by that time, will already be too late.

Writing for the *Pacific Ecologist* in 2002, Sharon Beder was already focusing on the newly-elected George W. Bush and his administration.[1] While many were shocked when the United States withdrew from the Kyoto agreement in March 2001, those familiar with the political climate merely viewed it as payback to oil and gas interests. One Democratic Congressional representative estimated that, during the 2000 election campaign, the coal industry contributed $3.8 million, with nearly 90 percent going to Republicans. The oil industry is estimated by the Center for Responsive Politics to have contributed $14 million with $10 million going to Republicans.

As the Kyoto Conference approached, the fossil fuel industries in the United States stepped up their campaign to prevent a treaty being signed that involved greenhouse gas reduction targets. A consortium of 20 organizations had launched an anti-climate treaty campaign in 1997. Industry groups representing oil, coal, and other fossil fuel-related interests, spent an estimated $13 million on television, newspaper, and radio advertising in the three months leading up to the conference to promote public opposition to the treaty. Speaking at a news conference during this campaign, Jerry Jasinowski, the president of the National Association of Manufacturers, argued that the treaty would result in higher energy prices, create job losses, and would harm businesses, farmers, and consumers.

---

1 Sharon Beder, *"Casting Doubt and Undermining Action,"* Pacific Ecologist, March 2002, pp. 42-49.

In 1998, the New York Times reported internal American Petroleum Institute documents showing that fossil fuel interests intended to raise $5 million over two years to establish a Global Climate Science Data Center as a non-profit educational foundation to help meet their goal of ensuring that the media and public recognize the uncertainties in climate science. The documents stated that victory would be achieved when climate change becomes a non-issue and when those promoting the Kyoto Protocol using existing science appear "to be out of touch with reality."

Beder quotes Phil Lesly, author of a handbook on public relations and communication, about the importance of fostering doubt in the minds of the public: "People generally do not favor action on a non-alarming situation when arguments seem to be balanced on both sides and there is a clear doubt. The weight of impressions on the public must be balanced so people will have doubts and lack motivation to take action. Accordingly, means are needed to get balancing information into the stream from sources that the public will find credible. There is no need for a clear-cut 'victory.'... Nurturing public doubts by demonstrating that this is not a clear-cut situation in support of the opponents usually is all that is necessary."

This strategy was also endorsed by Republican pollster Frank Luntz, even as the science documenting climate change was piling up.

The various corporate front groups formed during the 1990s is truly mind-boggling: an incomplete list would include the Global Climate Information Project, the Coalition for Vehicle Choice, the Advancement of Sound Science Coalition, the Information Council on the Environment, the Global Climate Coalition, the International Petroleum Industry Environmental Conservation Association, the Greening Earth Society, and the Center for the Study of Carbon Dioxide and Global Change. Others will be discussed below.

The Global Climate Coalition, originally a coalition of 50 trade associations and private companies in the United States representing oil, gas, coal, automotive, and chemical companies and trade associations, put together with the help of public relations giant Burson-Marsteller, spent millions of dollars in its campaign to persuade the public and governments that global warming was not a real threat. On its home page the Global Climate Coalition said its concern was with the "potentially enormous impact that improper resolution [of global climate issues] may have on our industrial base, our customers and their lifestyles, and the national economy."

But companies started to bail out after the GCC began to receive bad press in 1997. The first to leave was DuPont, followed by British Petroleum, Royal Dutch/Shell, Ford Motor Company, Daimler-Chrysler, Texaco, and The Southern Company. General Motors left in March 2000, three days after the National Oceanic and Atmospheric Administration published a report showing that the U.S. had experienced the warmest winter on record. By 2000, so many companies had left the GCC because of its poor reputation and the increasing evidence of climate change that the GCC had to restructure. The GCC also toned down its own public stance, arguing for voluntary measures to reduce emissions rather than disputing the need for mandatory measures.[1]

The Worldwatch Institute likened the exodus from the GCC to the demise of the Tobacco Institute, set up by the tobacco industry to undermine the certainty of the science that linked smoking with lung cancer and other diseases. The Tobacco Institute disbanded in 1999.

The Greening Earth Society was established in 1998 by the Western Fuels Association to convince people that "using fossil fuels to enable our economic activity is as natural as breathing."

Scientists who had been involved in the campaign to discredit emission-reduction targets included Richard Lindzen, Robert Balling, Sallie Baliunas, and S. Fred Singer. Dr. Lindzen, who had been listed as "an independent scientist," was a consultant to the fossil fuel industry, charging $2,500 a day for his services. Dr. Balling had also been heavily funded by fossil fuel interests. The *Arizona Republic* quoted him as saying that he had "received more like $700,000 over the past five years" from coal and oil interests in Great Britain, Germany, and the United States. Singer was executive director of the Science and Environmental Policy Project which argued that global warming, ozone depletion, and acid rain were not real but rather were scare tactics used by environmentalists. He had worked for Exxon, Shell, and Arco. According to the Environmental Research Foundation: "For years, Singer was a professor at the University of Virginia where he was funded by energy companies to pump out glossy pamphlets pooh-poohing climate change. Singer hasn't published original research on climate change in 20 years and is now an 'independent' consultant, who spends his time writing letters to the editor, and testifying before Congress, claiming that ozone-depletion and global warming aren't real problems."

Because these and a handful of other contrarian scientists had been trotted out so often by those trying to discredit the scientific consensus on global warming and because so much of the funding had been traced back to industry and their related associations, new "clean" scientists had to be found to replace these industry warhorses. An American Petroleum Institute memo revealed a new plan: "Identify, recruit and train a team of five independent scientists to participate in media outreach.... This team will consist of new faces who will add their voices to those recognized scientists who are already vocal."

Think tanks are generally private, tax-exempt, research institutes that present themselves as providing impartial expertise. However, these organizations

---

1 Lester R. Brown, *"The Rise and Fall of the Global Climate Coalition,"* Earth Policy Institute, July 25, 2000.

generally tailor their studies to suit clients or donors. These corporate-funded think tanks have played a key role in providing credible "experts" who dispute scientific claims of existing or impending environmental degradation and therefore provide enough doubts to ensure "lack of motivation" to act by either governments or the public. These dissident scientists, usually not atmospheric experts, have often argued that future warming will be slight and may have beneficial effects.

The Heritage Foundation is one of the largest and wealthiest think tanks in the United States. This organization receives massive media coverage in the U.S. and is very influential in politics, particularly with Republicans. In 1998, the foundation published a paper entitled "The Road to Kyoto: How the Global Climate Treaty Fosters Economic Impoverishment and Endangers U.S. Security." It began: "Chicken Little is back and the sky is falling. Or so suggests the Clinton Administration.... By championing the global warming treaty, the Administration seeks to pacify a vociferous lobby which frequently has made unsubstantiated predictions of environmental doom."

The Chicken Little analogy has worked extremely well over the years.[1]

In the 1999 edition of its *Environmental Briefing Book for Congressional Candidates*, the Competitive Enterprise Institute — another huge player in the global warming debate — argued that "the Kyoto Protocol is a costly, unworkable, and inappropriate policy to suppress energy use around the world" and that the U.S. Senate should reject it. It argued that the "scientific case for an international climate treaty has collapsed" and, anyway, "no one should worry about a modest warming, should it occur" because it is likely to result in beneficial impacts.

One of CEI's publications, *The True State of the Planet*, was partially funded by the Olin Foundation, created by Olin Chemical.[2] In it Robert Balling claimed that: "[The] scientific evidence argues against the existence of a greenhouse crisis, against the notion that realistic policies could achieve any meaningful climatic impact, and against the claim that we must act now if we are to reduce the greenhouse threat."

CEI is an active member of the Cooler Heads Coalition. This coalition was founded by the corporate front group Consumer Alert and distributed a biweekly newsletter, published by CEI. The object was clear: "The Cooler Heads Coalition focuses on the consumer impact of global warming policies that would drastically restrict energy use and raise costs for consumers."

Think tanks have been so successful at clouding the scientific picture of greenhouse warming and providing an excuse for corporations and the politicians they

---

1 Edwin J. Feulner, *"No Evidence for Global Warming,"* The Heritage Foundation, Jan. 22, 1998.

2 Ronald Bailey, *"The True State of the Planet: Ten of the World's Premier Environmental Researchers in a Major Challenge to the Environmental Movement,"* Competitive Enterprise Institute, Jan. 1, 1995.

support that they have, to date, managed to thwart effective greenhouse gas reduction strategies being implemented in the rest of the English-speaking world.

American think tanks also travel beyond U.S. borders to secure allies. In 2001 the Frontiers of Freedom Institute, a conservative corporate-funded U.S. think tank, organized a conference in Canberra, Australia, in conjunction with the Australian APEC Study Centre. The conference, "Countdown to Kyoto," was organized, according to *The Australian*, to "bolster support" for the Australian government's increasingly isolated position on global warming in preparation for the Kyoto Conference. Senator Chuck Hagel, who co-sponsored the Senate resolution condemning ratification of any treaty agreement in Kyoto which harmed the U.S. economy or failed to include commitments by developing nations, was a speaker as was U.S. Congressman John Dingell. Malcomb Wallop, who heads the Frontiers of Freedom Institute, chaired the conference. Wallop, who was a U.S. senator for 18 years, boasted of his achievements in promoting the Strategic Defense Initiative ("Star Wars") during the Reagan Administration and opposing welfare, progressive taxation, Social Security, and government funding for higher education. Wallop said in a letter to U.S. conservative groups: "This conference is a shot across the bow of those who expect to champion the Kyoto treaty." He added that the conference would "offer world leaders the tools to break with the Kyoto treaty."

Patrick Michaels, who has been funded by U.S. coal interests, has traveled the world on behalf of anti-climate treaty interests. In 1997, he attended a conference similar to the Australian conference in Vancouver organized by a Canadian conservative think tank, The Fraser Institute. Also attending this conference were Robert Balling and Sallie Baliunas.

Many multinational energy companies have adjusted their advertising to put on a "greener" face. In 1997, when it left the Global Climate Coalition, BP (what once stood for "British Petroleum" was at the moment defined as "Beyond Petroleum," until BP executives dropped the initiative when it didn't pass the laugh test) earned the reputation of being environmentally progressive in an industry that largely refused to accept that global warming was likely to occur. Before its conversion to environmentalism, BP was receiving bad publicity and criticism from human rights groups because of its activities in Colombia. In 1997, a year in which BP had adverse publicity about its activities in Colombia and favorable publicity about its stance on global warming, BP's share price and profits were up.[1]

Certainly BP's record of environmental protection has not been any better than other oil companies. It had contributed more than its share of oil spills and pollution. BP was cited as the most polluting company in the U.S. in 1991, based on the EPA's toxic release inventory. One local residents' group claimed: "BP has

---

1 Robert Verkaik, *"BP Pays Out Millions to Colombian Farmers,"* The Independent [U.K.], July 22, 2006.

treated us as a PR problem instead of taking our concerns seriously." In 1992, Greenpeace International named BP as one of Scotland's two largest polluters. BP was also named as "a potentially responsible party for 23 hazardous waste Superfund sites in the United States."[1]

Nor did BP become a model after its apparent environmental conversion in 1997. In 1999, BP was charged with burning polluted gases at its Ohio refinery and agreed to pay a $1.7 million fine. In July 2000, BP paid a $10 million fine to the EPA and agreed to reduce the air pollution from its U.S. refineries by tens of thousands of tons. The agreement, although voluntary, was taken to head off EPA enforcement action. In return for the agreement, the EPA agreed to a "clean slate" for certain past violations.

By 1999, BP's emissions were "greater than those of Central America, Canada or Britain." BP is now responsible for about 3 percent of worldwide greenhouse emissions, a huge amount for a single company.

Greenpeace USA gave BP CEO John Browne an award for "Best Impression of an Environmentalist." The environmental group noted that BP planned to spend $5 billion on exploring for and producing oil in Alaska even though the planet could "not afford to burn 75 percent of the fossil fuels we've already discovered, if we are to avert catastrophic global warming."

BP has now joined the Safe Climate, Sound Business Coalition which recommends ways to reduce greenhouse gas emissions without reducing economic growth. It affirms its intention of meeting "growing needs for energy" and emphasizes the need for "gradual transition" to "avoid the need to retire existing equipment prematurely."

"Actions that seek, at a stroke, drastically to restrict carbon emissions or even to ban the use of fossil fuels would be unsustainable because they would crash into the realities of economic growth," according to Browne.

"The public conversion of fossil fuel companies to mainstream greenhouse science is no real indication that they are committed to contributing to the reduction of global greenhouse emissions," wrote Beder, a professor at the University of Wollongong in Australia, "particularly where that interferes with their profits."

Professor Beder offers the following conclusion:

> Clearly, corporations and industries that depend on fossil fuels for the greater part of their profits are doing their best to obstruct and undermine effective measures to reduce greenhouse gases. The front groups, scientists, economists and think tanks on their payroll need to be exposed for what they are — the voices of vested interests. In addition, the voice of the public needs to drown out the influence of corporations, their lobbyists, advertisements, and their political donations. Unless politicians are swamped by protests, letters and phone calls demanding action to prevent global warming, it will be too easy for governments to accede to corporate interests.

---

1 *"Pick Your Poison,"* Sierra Magazine, The Sierra Club, Sept.-Oct. 2001.

Writer Michael Shnayerson took an extensive look at one think tank, the Competitive Enterprise Institute, in an article for *Vanity Fair* in May, 2007.[1] The story focuses particularly on Myron Ebell, the director of Global Warming and International Environmental Policy at the CEI. He is also the chairman of the Cooler Heads Coalition. The media often turn to Ebell for the "opposing view" on climate change. As Shnayerson points out, journalists like to air all views — "on the one hand, on the other" — so his opinions are given equal weight with new scientific findings. Ebell is not a scientist.

"Like holdouts in the Alamo, the last of the skeptics plug away at the thousands of mainstream scientists now arrayed against them," Shnayerson writes. "They take potshots at the scores of studies that say global warming is here, aiming for small incongruities. And they bridle when asked if they take money, as nearly all do, from ExxonMobil."

Ebell's job is to convince the media and the public that a controversy is still raging. Journalist Ross Gelbspan said that skeptics operate by planting doubts in the press causing the public to shrug, politicians to push the problem off to another day, and ExxonMobil to fight proposed regulations, earning windfall profits in exchange for the pittance paid to their hired contrarians.

In February 2008, ExxonMobil announced their gangbuster record earnings of $11.7 billion, setting a U.S. record for the biggest quarterly profit, for the final three months of 2007. This beats its own mark of $10.71 billion set in the fourth quarter of 2005. Exxon's annual profit in 2006 was $39.5 billion. Exxon broke its own profit record again when figures were released in 2009.

In 2005, ExxonMobil gave CEI $270,000 and between 1998 and 2005 ponied up more than $2 million. The think tank also received money from the American Petroleum Institute, various pharmaceutical companies (Dow Chemical, Eli Lilly), and William A. Dunn of Dunn Capital Management. CEI was founded 23 years ago by free marketeer Fred Smith to fight government regulation and current issues have included opposing higher mileage standards in cars and the Endangered Species Act.

"Other oil companies fund global-warming-skeptic think tanks through the American Petroleum Institute, and various coal interests weigh in, too. But, for the skeptics, ExxonMobil is Big Daddy," according to Shnayerson. He reported that the giant oil company spent $16 million funding climate studies at some three dozen institutes from 1998 to 2005. None of these groups, he points out, had any standing in mainstream climate science. Their papers were taken up by politicians like James Inhofe and Texas Republican Joe Barton and became widely disseminated on the Internet. A trip to Internetland will clearly demonstrate the effectiveness of this approach.

---

1 Michael Shnayerson, "*A Convenient Untruth*," Vanity Fair magazine, May 2007.

"The case for global warming has grown all but irrefutable, yet the skeptics have enjoyed enormous influence, for the audience that matters most to them occupies the White House," Shnayerson wrote during the George W. Bush Administration. "Eagerly, their papers have been snatched up by the Bush Administration as rationales for all manner of public policy, from striking down the Kyoto Protocol to blocking any cap on carbon dioxide emissions."

CEI ran a public service commercial on television with the tagline: "They call it pollution. We call it life."

"Ebell actually seems to believe what he's saying. Which is remarkable, really, because every one of his arguments, put to scrutiny by a murderer's row of the country's top climate scientists, seems to fall apart," Shnayerson wrote.

Forced to respond to Ebell's comments because of wide media play, mainstream climate scientists seemed frustrated to the point of apoplexy. The House of Commons in the United Kingdom proposed a motion to censure Ebell. Tom Wigley, a senior scientist at the National Center for Atmospheric Research, responded thusly to Ebell's analysis on ocean warming: "That's the most preposterous bullshit I've ever heard." According to Shnayerson, Ebell apparently revels in upsetting these experts.

David Legates is another hard-core skeptic, director of the University of Delaware's Center for Climatic Research, who issued a paper declaring that only two of the world's 20 polar bear populations were decreasing. Most of the others are stable; two are growing. Legates' paper was published by another think tank, the National Center for Policy Analysis — whose global-warming-denial research was partially funded in 2005 with a $75,000 contribution from ExxonMobil — not in a peer-reviewed scientific journal.

While Ebell likes to bash scientists for working outside their degreed fields, Ebell, as noted earlier, is not a scientist at all. He majored in philosophy at the University of California in San Diego, and then studied political theory at the London School of Economics and history at Cambridge.

"Every interview I do, when I'm asked about scientific issues, I say I'm not a climate scientist," Ebell told *Vanity Fair*. "I'm just giving you the informed layman's perspective.... If science is going to be discussed in the public arena, then shouldn't people other than scientists be allowed to participate? Isn't that what representative democracy is?"

Some of Ebell's observations exhibit such a simplistic view of global warming that some found it hard not to believe that they weren't a put-on.

"People prefer warmer climates," Ebell told *Vanity Fair*. "They do better in them. People do better in Phoenix than they do in Buffalo. They feel better, they're happier, they're more productive. They live longer." Ebell, by the way, lives in a suburb of Washington, D.C.

Ebell, however, could be right about one thing. He said that if Al Gore is correct and emissions will have to be reduced by 60 to 80 percent by 2050, the goal will never be met.

"Between now and 2030 — not 2050 — world energy use is going to go up by 60 percent," Ebell said. "This (emissions reduction) is a vision coming up against reality. It's a train wreck, and I know which side will win. Reality will win."

So while Ebell pooh-poohs climate change, he also seems to be saying that any attempt to stem the problem is hopeless. Emissions will increase, the situation will worsen, and we will all live happily in Phoenix.

"Ebell may be proved right about Kyoto," longtime *New York Times* environmental reporter Andrew Revkin told Shnayerson. "And because global warming isn't a catastrophe in the traditional definition of the word, its full impact won't be felt tomorrow, or next year, or perhaps even for decades. That makes it easy for Ebell to target some scientists as alarmists. This can perpetuate a false sense of intellectual deadlock. And that masks what is clearly established — that more carbon dioxide will raise temperatures and seas for centuries to come."

ExxonMobil stopped funding CEI in 2006.

The true effects of global warming and its effects on social, political, and economic institutions won't become evident for decades. Think lung cancer. The effects of smoking aren't immediate until decades after someone picks up a cigarette. The Earth has been smoking since the Industrial Revolution and has become a two-pack-a-day man since the 1970s. The Earth will bump that up another pack by 2030 following the "business-as-usual" path and the sickness will be obvious as well as unstoppable.

The denial machine, as noted repeatedly, has been very effective in influencing the American public and successfully delaying any serious action on climate change. The machine is still operating effectively and the clock is still ticking. The deniers find a very receptive audience for myriad reasons. People are busy. People don't want to face the possibility of change. People want better lives than their parents and to keep up with their peers. The country is struggling; people are struggling. But other factors are also at work.

The denial machine has equated opposition to change with personal freedom. What should be a common concern has become political with deniers slamming the old straw men of big government and pointy-headed intellectuals intent on telling people how to live their lives. The absurdity of the situation is that people who consider themselves fiercely individualistic have been led around by the nose for the sake of corporate profits, often at the expense of their own jobs and quality of life. Recent polls have shown that an overwhelming majority of Americans now consider climate change a major, even urgent, problem. Yet global warming ranks well down the list of voter concerns. While people may be concerned about climate change, poll numbers drop into the 20s when taxpayers are asked to fund the changes necessary to adjust to the new and future realities.

Some years ago, when CAFE standards were being considered for the umpteenth time, a C-Span caller summed up a view held by many Americans. "Nobody's goin' to tell me what kind of goddamn car or truck I can drive. I got the right to do whatever the hell I want." Freedom, then, means the right to do something regardless of the consequences to others. Car dealerships in Washington could not keep the Cadillac Escalades in stock as these behemoths became the cars of choice for former Bush Administration officials. The message was clear: screw you, we do what we want. You can, too. That's freedom.

Fortunately for the denial machine, most Americans simply aren't interested in science. For example:

In a 1996 Associated Press poll, less than half of American adults understood that the Earth orbits the sun yearly.

In a Fox News/Opinion Dynamics survey in 1996, 1 percent of respondents cited the environment as the country's most pressing issue.

Less than half of the American population accepts the theory of evolution. In a 2004 survey, more than half agreed that "we depend too much on science and not enough on faith."

More people in the United States believe that a house can be haunted than accept the theory of the Big Bang, and 29 percent are not certain that the Earth revolves around the sun rather than vice versa, according to the National Science Foundation.

The combination of religious worldviews and beliefs in pseudo-science (astrology, UFOs, etc.) would encompass more than half of the American population. These belief systems set America apart from other advanced nations. This ignorance of the natural world does not bode well for the nation when it comes to substantial environmental policy change, especially in light of the well-funded public relations campaigns of the fossil fuel industries and staunchly free-market think tanks.

Tom Henry, an environmental writer for the *Toledo Blade* in Ohio, has come to expect a barrage of hate mail whenever he even mentions global warming. Ignoring climate change would be difficult since his beat includes Lake Erie and the rest of the Great Lakes.

"The reasons for hostility are varied," Henry said. "Some are convinced this issue can be blamed on a bunch of tree huggers who got together one day and created a bunch of fear-mongering in hopes of bringing energy production to a halt and wrecking the U.S. economy; despite the fact legitimate mainstream scientists have been studying this for more than 30 years.

"Some are convinced it's all propaganda being pushed by a bunch of wacky, far-left Democrats even though George W. Bush himself — quite possibly the worst environmental president ever — has repeatedly gone on record since his first year in office and acknowledged that even he is now convinced climate change is happening. He just doesn't agree with the Kyoto Protocol as a vehicle for curbing emissions.

"Everyone says they want the issue based on science, not politics (as if you can separate the two). Well, science has spoken — it's a slam dunk. Any scientist still raising doubts about climate change needs to have his/her credentials/funders examined....

"Yet some people only will believe what they want to believe. They will cite authors such as Bjorn Lomborg and fiction writer Michael Crichton, both of whose works have been widely discredited, while ignoring the IPCC and its 1,500 scientists, including the world's top climatologists.

"Are there extenuating circumstances? Of course. Yes, indeed, there are natural fluctuations in climate. The sun goes through periods of cooling and warming. Volcanic activity put an incredible amount of carbon dioxide into the atmosphere. There would be some climate change even if man never invented the automobile or heated a single home or invented electricity. The question is how much is man-made versus natural and what can we do about it to slow down or mitigate the effects.

"One of the Great Lake's region's top climate scientists once told me roughly two-thirds of climate change is natural and one-third is man-made — but everything depends on how far you're going to let those man-made emissions go. But, to some people, as soon as they hear *anything* can be attributed to natural changes, they immediately use that as justification to dismiss man-made activity and assume they can carry on with their antiquated, polluting ways.

"I've heard several people say how they heard in the 1960s or 1970s we were heading into an ice age, not a warming period. So they dismiss everything they've heard over the past 30 to 40 years. I think it's obvious our technology and expertise is better now than it was in the 1960s or 1970s, so I don't get that argument, either.

"Many don't realize that some parts of the Earth actually *are* expected to be cooler while the poles melt. That's how the term 'climate change' overtook global warming, because there will be climate variances, such as more violent storms. Try getting people to understand there will be more rain in the future but less water because evaporation will outpace your gains in precipitation. It's a hard sell.

"The bottom line is that there's a lot of confusion and a lot of money at stake. And when that happens, some people believe only what they want to believe."

However, 53 percent of survey respondents viewed environmental protection as more important than economic growth. But that was before the current recession and climate change all but dropped off the political radar. So the denial machine is still engaged in a tight race. Many skeptics believe — and quite possibly rightly so — that an economically stressed public will not devote the political will necessary to make appropriate changes.

A good example of commentary found on the Internet is a column written by one Deneen Borelli titled "Global Warming Statists Threaten Our Liberty." She has no scientific credentials. Her biography says she worked for Philip Morris and as a runway fashion model, and "auditioned for television commercials." Her hobbies include pistol target shooting.

"Life, liberty and the pursuit of happiness — 'unalienable [sic] rights' cited by our Founding Fathers in the Declaration of Independence — are now at risk as left-wing activists seek to curtail our liberties and personal choices to save the planet from supposedly man-made global warming."

Borelli goes on to complain about restrictions on SUVs, light bulbs, food, and other assaults on her consumption which she considers her "unalienable" right.

"Despite the numerous flaws and ambiguities in trying to link human behavior and global warming, activists and their allies in government use emotion and alarmism to make their case. They are seeking to cut off any responsable [sic] debate and silence their critics by saying these people are motivated by corporate and personal greed and don't care about pollution. That, however, is hardly the case.

"Critics of the global warming agenda are motivated instead by a love of freedom and civil liberties. They want a discussion based on logic and facts that will address any problems without depriving us of liberty and personal choice. They do not want to sacrifice our way of life based on fears of an unproven theory. After all, the loss of liberty is a greater cause of alarm than global warming."

Borelli's column was published by The National Center for Public Policy Research, a conservative think tank founded in 1982. "We believe the principles of a free market, individual liberty, and personal responsibility, combined with a commitment to strong national defense, provide the greatest hope for meeting the challenges facing America in the 21st century," according to their website. The organization received $45,000 from ExxonMobil in 2002 and $55,000 in 2003. One published goal is to combat information disseminated by the "environmental left."

President Amy Ridenour has held the post since its founding and her husband, David, is vice president. Both were active members of the College Republican National Committee.

NCPPR bulletins have been heavily cited by conservative members of Congress, sometimes apparently being copied entirely with little more than a change in letterhead. Ms. Ridenour is pro-tobacco and anti-environment. She has also been closely linked to convicted lobbyist Jack Abramoff and attended a 1997 Tom

DeLay-Abramoff trip to Russia funded through the NCPPR by the Russian energy giant Naftasib.[1]

"It's been estimated that for every $5 billion to $11.5 billion in regulatory costs, one person dies prematurely each year," David Ridenour has been quoted as saying. "Assuming it cost $200 billion per year to stabilize our greenhouse gas emissions, between 17,000 and 40,000 Americans could die prematurely. Sadly, such a sacrifice would be made despite the fact that satellite and weather balloon data indicate that the temperature of the planet has been declining over the past 18 years."

The NCPPR has been a loyal conservative Republican ally as climate science has become more and more politicized. Politicians are good at politics. Scientists aren't. Climate change poses the biggest threat to unrestrained free-market capitalism and globalization since the Soviet era. Eminent scientists are questioning the wisdom — even the possibility — of unrestrained free-market global growth. On the right side of the political divide, the "invisible hand" of the free market has attained the status of a secular religion with huge payoffs for the devotees. The clash was inevitable and the stakes are enormous. Global warming threatens to halt the juggernaut of American free market economics and any change will be fought tooth-and-nail by corporate interests and free-market ideologues.

Groups like the NCPPR have been good soldiers.

"I applaud the NCPPR staff and supporters for your commitment to educating Americans on ... issues that are essential to our nation's prosperity and security," wrote President George W. Bush. "For 20 years, you have remained steadfast in your efforts to advance the cause of individual freedom in the United States."

Talk show host Rush Limbaugh has said, "I think those guys are brave and courageous.... They've got guts."

The think tank said it has been particularly successful "in today's competitive media environment, garnering over 4,996 media interviews and citations in 2006...." Not surprisingly, NCPPR representatives have appeared on talk radio shows with Limbaugh, Michael Reagan, G. Gordon Liddy, Sean Hannity, Roger Hedgecock, Glen Beck, "and many others."

One group worth a special look is the libertarian Ludwig von Mises Institute in Auburn, Ala. The president and founder is Lew Rockwell, whose columns are widely published on the Internet. Environmentalists are portrayed by Rockwell as, well, murderers. He writes:

"The politics of the environmentalists are increasingly predictable and obvious. They oppose all forms of capitalistic innovation. Indeed, they represent a special kind of danger to the human race that socialism never did. The greens are

---

1   James V. Grimaldi and Susan Schmidt, *"Report Says Nonprofits Sold Influence to Abramoff,"* Washington Post, Oct. 13, 2006.

against all that. They claim that we should be happy to live amidst disease, filth, and death, if only the bugs and birds can be left alone to thrive and kill us.

"It's as if the socialists discovered that their plan created poverty, and so decided to change their names to environmentalists and make poverty their goal.

"And note how their agenda fits so well with the state agenda. The state hobbles and hinders productivity in millions of ways through its taxing, regulating, and warmongering. But that makes little difference to the state, which prefers the exercise of power to the good of society. So too do the environmentalists pursue their agenda without regard to the effects on human society.

"Currently, there are many environmental issues alive in the policy world, from the debate over sprawl to the frenzy over global warming. The environmentalists have the upper hand in all of them, which is a crying shame, given that they're responsible for millions of lost lives — in just one of their conquests. More victories for them are sure to make life worse for all of us."

And Rockwell is still fighting the good fight against long-dead Rachel Carson with his Internet diatribes.

"[DDT] was used throughout the 1950s and 60s and was on the verge of wiping out mosquito-borne diseases from the planet," Rockwell wrote in 2006. "Then something very peculiar came along. A book called *Silent Spring* by Rachel Carson was published in 1962, and eventually it created a fantastic backlash against progress. The spring was silent supposedly because of the lack of birds, killed off by DDT.

"The only problem is that Carson's claims were never scientifically validated. Indeed, it was a hoax. Studies pumped primates full of DDT with no effect. Human volunteers ingested the stuff with no effect. Workers with 600 times the typical exposure to DDT showed no increased side effect. ... What's more, she never once mentioned in her book that DDT had saved hundreds of thousands of lives."[1]

Rockwell, of course, is not a scientist. He has a degree in English from Tufts University and is very close to fellow libertarian and unsuccessful presidential candidate Ron Paul.

The Cato Institute is a libertarian think tank founded in 1977 by Edward H. Crane. The institute is named for Cato's Letters, a series of libertarian pamphlets that the institution contends helped to lay the philosophical foundation for the American Revolution. According to its mission statement, the institute "seeks to broaden the parameters of public policy debate to allow consideration of traditional American principles of limited government, individual liberty, free markets, and peace. Toward that goal, the institute strives to achieve greater involvement of the intelligent, concerned lay public in questions of policy and the proper role of government."

1 Llewellyn H. Rockwell, Jr., *"Our Own Silent Spring,"* LewRockwell.com, June 29, 2006.

To its credit, Cato tries to limit corporate money. Between 1998 and 2004, the Cato Institute did received $90,000 of its funding from ExxonMobil — about a tenth of a percent of the organization's budget over that period. Cato has also received funding from the American Petroleum Institute.[1]

Cato scholars have written extensively about environmental issues, including global warming, environmental regulation, and energy policy. The institute's work on global warming has been a particular source of controversy. The institute has held a number of briefings on global warming with global warming skeptics as panelists. In December 2003, panelists included Patrick Michaels, Robert Balling, and John Christy. Balling and Christy have since made statements indicating that global warming is, in fact, related at least to some degree of human activity.

In response to the *World Watch Report* in May 2003 that linked climate change and severe weather events, Jerry Taylor of Cato said: "It's false. There is absolutely no evidence that extreme weather events are on the increase. None. The argument that more and more dollar damages accrue is a reflection of the greater amount of wealth we've created."

Four out of five "Doubters of Global Warming" interviewed recently by PBS's *Frontline* were funded by, or had some other institutional connection with, the institute. Cato has often criticized Al Gore's stances on the issue of global warming and agreed with the former George W. Bush Administration's skeptical attitude toward the Kyoto Protocols.

On the other hand, Cato scholars had been critical of the former Bush Administration's views on energy policy. In 2003, Cato scholars Taylor and Peter Van Doren blasted the Republican Energy Bill as "hundreds of pages of corporate welfare, symbolic gestures, empty promises, and pork-barrel projects." They also spoke out against Bush's call for larger ethanol subsidies.

In 2006, a 700-page assessment of global impacts of climate change was released. The *Stern Review on the Economics of Climate Change* was conducted by Sir Nicholas Stern, head of the Government Economic Services in the United Kingdom and former World Bank economist. The review concluded that urgent and international action was required to avoid the worst impacts of climate change and urged cooperation between governments, businesses, and individuals in addressing the challenge. The Royal Society submitted evidence for the report in December 2005. According to the Society, "it acknowledged the need for strong and deliberate government policy choices to motivate change, and warned that there was no time for delay."[2]

Royal Society President Martin Rees, who chaired the launch proceedings, said, "The Stern Review emphatically points to the need to take prompt and

---

1 For a complete list of Cato Institute donors, see Fact Sheet: Cato Institute, ExxonSecrets.org.
2 "*Stern Review on the Economics of Climate Change*," The Royal Society, Excellence in Science — archive 1, royalsociety. org, Dec. 2006, London, England.

strong action to avoid the worst economic and environmental costs of climate change. This should be a turning point in a debate which has pitted short-term economic interest against long-term costs to the environment, society, and the economy."

Cato analyst Indur Goklany responded to the report in February 2008. The executive summary states that the Stern report really "indicates that through the year 2100, the contribution of climate change to human health and environmental threats will generally be overshadowed by factors not related to climate change. Hence, climate change is unlikely to be the world's most important environmental problem of the 21st century."

Goklany concluded that "equity-based arguments, which hold that present generations should divert scarce resources from today's urgent problems to solve potential problems of tomorrow's wealthier generations, are unpersuasive."

According to Goklany, "Halting climate change would reduce cumulative mortality from various climate-sensitive threats, namely, hunger, malaria, and coastal flooding, by 4 to 10 percent in 2085, while increasing populations at risk from water stress and possibly worsening matters for biodiversity. But according to cost information from the U.N. Millennium Program and the IPCC, measures focused specifically on reducing vulnerability to these threats would reduce cumulative mortality from these risks by 50 to 75 percent at a fraction of the cost of reducing greenhouse gases. Simultaneously, such measures would reduce major hurdles to the developing world's sustainable economic development, the lack of which is why it is most vulnerable to climate change.

"The world can best combat climate change and advance well-being, particularly of the world's most vulnerable populations, by reducing present-day vulnerabilities to climate-sensitive problems that could be exacerbated by climate change rather than through overly aggressive greenhouse gas reduction."[1]

If this explanation seems dubious and confusing, it is. The goal is to invest in solving water and biodiversity problems caused by greenhouse emissions while not addressing the primary problem that caused the crisis in the first place. The important issue for Cato: no government regulations on greenhouse gas emissions regardless of the consequences. Any admission that government regulation would be needed to stem greenhouse gas emissions would send the libertarian ideology crumbling down like a house of cards.

The Acton Institute in Grand Rapids, Mich., has managed to meld libertarianism with Christianity. The institute is notable because it embraces both unregulated free markets and biblically-based views of man's role in nature and, consequently, climate change.

---

1 Indur Goklany, *"What to Do about Climate Change,"* The Cato Institute, Policy Analysis no. 609, Feb. 5, 2008.

The institute received $50,000 in 2007 from ExxonMobil, bringing the institute's total received from ExxonMobil to $215,000, according to Internal Revenue Service documents and Exxon records.[1] While receiving less money and attention than some of the larger think tanks funded by ExxonMobil, the Acton Institute advances views and policies friendly to the energy giant. The institute brought global warming skeptic Fred Smith of the Competitive Enterprise Institute to Grand Rapids and also began running a regular blog feature dedicated to challenging the idea that there is scientific consensus on climate change.

Some insight into the thrust of the Acton Institute's views on global warming can be gleaned from a commentary written by Jay W. Richards:

> These [biblical] truths provide a solid theological foundation for addressing environmental concerns while avoiding an anti-human bias. Unfortunately, these truths do not figure prominently in the contemporary debate. In fact, it's more fashionable to argue — incorrectly — that the Judeo-Christian tradition is the problem, not the solution. Even some Christians who have entered the fray have not been careful to separate the empirical evidence from the doubtful assumptions.
>
> An organization called the Interfaith Stewardship Alliance has been launched to help Jews and Christians develop a positive environmental ethic that avoids such pitfalls. Announced [in 2005] at the Ugandan embassy in Washington D.C., the ISA is a coalition of individuals and institutions — including the Acton Institute — who share an interest in environmental stewardship....
>
> The ISA draws its inspiration from the Cornwall Declaration, published by the Acton Institute in 2000. As theologian Calvin Beisner explains, the Cornwall Declaration describes human beings not merely as consumers and polluters but also producers and stewards. It challenges the popular assumption that 'nature knows best,' or that 'the Earth, untouched by human hands, is the ideal.' And it calls for thoughtful people to distinguish environmental concerns that 'are well-founded and serious,' from others that 'are without foundation or greatly exaggerated.' In other words, it calls for a reasoned, humane environmental ethic. At a time when mistaken policies based on anti-human assumptions can lead to the deaths of millions of people, such an ethic cannot come soon enough.[2]

Additionally, the Acton Institute views population growth as a good thing — flying in the face of any serious research done by demographers.

In 2006, the Cornwall Alliance (formerly known as the Interfaith Stewardship Alliance) published *A Call to Truth, Prudence, and Protection of the Poor: An Evan-*

---

gelical Response to Global Warming.[1] The alliance describes itself as a coalition of "religious leaders, clergy, theologians, scientists, academics, and other policy experts committed to bringing a proper and balanced biblical view of stewardship to the critical issues of environment and development."

The 19-page document is a critique of another paper, *Climate Change: An Evangelical Call to Action,* published earlier, that called for Christian evangelicals to take a more significant role in climate change issues.[2]

The difficulties for evangelicals in addressing climate change revolve around a number of issues: primarily, only God can have a significant impact on man and nature; man, being God's most important creation, is intrinsically "good"; all nature exists to serve man's needs and man is the steward and master of the planet; and "to increase and multiply" is biblically sanctioned. Science recognizes that man has had a negative impact on natural systems; that nature has its own rules to which man is subject; and overpopulation is a serious issue that will have to be addressed in the 21st century. To evangelicals and other religious groups, God is in control and the belief that man is in any way responsible for the troubles on God's Earth is simply hubris.

If religious fundamentalists accept that the Bible has erred about the age of the planet, the nature of the universe, the role of evolution and other issues, this would undermine what are believed to be other inerrant biblical views. And if the Bible is wrong on some fronts, why couldn't it also be wrong about other issues? Climate change has become a litmus test about the literal truth of the Bible. The end result has been an uneasy alliance of corporate interests, conservatives, libertarians, and evangelicals. The sheer number of people belonging to one of these communities and interest groups would easily be more than half of the American population. In polls, evangelicals have shown that they are much less concerned about climate change than any other group.

The Cornwall Alliance paper is not an attempt to reconcile biblical teachings with climate science — it's an attempt to debunk the science. No compromise is possible. These fundamentalists are also a big factor in the denial machine.

The most important paragraph comes at the very end of what the Cornwall Alliance tries to present as a scientifically-based paper: "Today we stand with the Oxford Declaration in deploring policies, laws, and regulations whose effect is to favor the already wealthy at the expense of the still poor, excluding them from legitimate participation in advanced economies and all the benefits they deliver such as lower infant child mortality rates, longer life expectancy, lower disease rates, more and better education, transportation, communication, and all

---

1   E. Calvin Beisner *et al., "A Call to Truth, Prudence, and Protection of the Poor: An Evangelical Response to Global Warming,"* Cornwall Alliance For The Stewardship Of Creation, www. CornwallAlliance.org, 2006.

2   Laurie Goodstein, "Evangelical Leaders Join Global Warming Initiative," The New York Times, Feb. 8, 2006.

the other things the already wealthy take for granted. Therefore we pledge to oppose quixotic attempts to reduce global warming. Instead, constrained by the love of Jesus Christ for the least of these (Matthew 25:45), and by the evidence presented above, we vow to teach and act on the truths communicated here for the benefit of all our neighbors."

The driving force behind the Cornwall Alliance is Dr. E. Calvin Beisner who, of course, is not a scientist. And the honorific of "doctor" comes from his earning a PhD in theology. Beisner is an associate professor of "social ethics" at Knox Theological Seminary in Florida. According to the Cornwall Alliance, he "lays a biblical foundation for the moral approach advocated by [Paul] Driessen and the prudent scientific caution advised by [Roy] Spencer. It is imperative, Beisner says, that Christians "make sure of their biblical moorings before venturing too far in the endorsement of specific policies, particularly when the policies can have serious consequences for human life and well-being."

"Our wise Creator has built multiple self-protecting and self-correcting layers into His world," contends Beisner, "which we have been given to use for our benefit as responsible environmental stewards. In deciding how to manage the Earth and its resources, the Bible requires that we consider the consequences of our actions — for wildlife, our planet, and the poorest among us."

Driessen suggests that a better approach to any problems would be "to develop technologies that generate more energy, at a lower cost and with fewer emissions — and export that technology to poor countries."

Roy Spencer is principal research scientist for the University of Alabama.

"Because of our limited understanding," says Spencer, "we cannot model or predict future climate cycles with any confidence. However, there is strong evidence that the Earth's natural 'greenhouse effect' acts like a blanket, working in conjunction with weather and hydrologic cycles to ensure long-term and global averages, despite local and short-term variations."[1]

Driessen concludes that higher carbon dioxide levels would promote longer growing seasons that would benefit plant growth and "actions to reduce energy use would adversely affect economic growth, human health, and societal well-being, while doing little to affect our climate."[2]

The Cornwall report, presented as an academic paper and duly footnoting the most notable skeptics on climate change, relies on research done by think tanks such as the aforementioned Cato Institute and the Competitive Enterprise Institute.

---

1 Roy Spencer, "*An Examination of the Scientific, Ethical and Theological Implications of Climate Change Policy, Executive Summary,*" The Cornwall Alliance, Implications of Climate Change Policy, Jan. 8, 2008, p. 1.
2 Driessen, ibid., p. 2.

According to the report, "climate variations are too poorly understood to be included accurately in computer climate models. Hence, the models risk overstating human influence."

Some of the conclusions are as follows:

• The executive summary of the IPCC "does not reflect the depth of scientific uncertainty embodied in the report and was written by government negotiators, not the scientific panel itself."

• Recently 60 topic-qualified scientists asserted that "global climate changes all the time due to natural causes and the human impact still remains impossible to distinguish from this natural noise," and that "observational evidence does not support today's computer climate models, so there is little reason to trust model predictions of the future."

• Because CO_2-induced warming will occur mostly in winter, mostly in polar regions, and mostly at night. But in polar regions, where winter night temperatures range far below freezing, an increase of 5.4 degrees Fahrenheit is hardly likely to cause significant melting of polar ice caps or other problems."

•"Even if the recent strong warming trend (at most 1 degree Fahrenheit in the last 30 years) is entirely manmade (and it almost certainly is not), and even if it continues for another 30 years (as it might), global average temperature will only be at most 1 degree Fahrenheit warmer than now. Predicting climate beyond then depends on assumptions about future use of fossil fuels. Such assumptions are dubious in light of continuous changes in energy sources throughout modern history. Who could have predicted our current mix of energy sources a century-and-a-half ago, when wood, coal, and whale oil were the most important components and petroleum and natural gas were barely in use?"

• According to a Competitive Enterprise Institute paper, "Of the costs to the Netherlands, Bangladesh, and various Pacific islands [i.e., the places at greatest risk], the costs of adapting to the changes in sea level are trivial compared with the costs of a global limitation of CO_2 emissions to prevent global warming."

• The death rate from severe cold is nearly 10 times as high as that from severe heat, implying that global warming (assuming that it reduces cold snaps as much as it increases heat waves) should prevent more deaths from cold than it causes from heat.

• A further implication is that because energy is a crucial component of economic development, affordable energy is necessary to protect against heat waves.

• W]hile worldwide data are insufficient to justify any generalizations, we do know that there is no statistical correlation between global average temperatures and droughts in the southwestern United States or even the United

States as a whole, a fact that puts model forecasts into doubt. Further, in an increasingly wealthy world, the ability to distribute water and agricultural products efficiently will continue to improve, making societies more and more resilient to droughts — which will continue to occur with or without human influence on climate.

• The impacts of climate change on malaria, at least through 2085, will be trivial compared to non-climate change-related factors.

• Observational evidence and computer models yield little confidence in forecasts of the impact of global warming on agricultural production, whether in poor countries or elsewhere. However, rising CO_2 — presumably what drives global warming — enhances agricultural yield. For every doubling of atmospheric CO_2 concentration, there is an average 35 percent increase in plant growth efficiency. Plants grow better in warmer and colder temperatures and in drier and wetter conditions, and they are more resistant to diseases and pests. Consequently their ranges and yields increase. Agricultural productivity worldwide and in developing countries has never been higher than it is today. Three likely results of rising CO_2 are shrinking deserts, lower food prices, and reduced demand for agricultural land to feed the world's population, the latter resulting in reduced pressure on habitat and consequently on species survival. These benefits would be reduced or foregone if we reduced atmospheric CO_2.

• The IPCC summary "was politically driven." They "produce scenarios with no basis in actual evidence. They are based on grossly unrealistic assumptions about future energy use, dominant energy types, pollution levels, economic development, and other factors that do not reflect current facts or likely future situations. Mainstream media generally report on worst-case scenarios and assume that warming will be catastrophic and will bring devastating harm but no benefits."

• There is evidence that the current warming period, from the mid-1880s to the present and likely to continue for a century or more, is driven largely by natural causes.

• The popular belief that there is such a consensus [on human-induced global warming] is dubious at best.

• Solar variability is a key driver of recent climate change; and ... climate modeling is highly uncertain.

• The idea of scientific consensus on anthropogenic global warming is an illusion.

• It is ironic that many supporters of the ECI rely heavily on the claim of scientific consensus to buttress their view of global warming. The role of the IPCC in climate studies is similar to that of the Jesus Seminar in New Testament scholarship in the 1990s and Darwinism for the past century. It is a self-selecting group with a narrow point of view favored by the political left and mainstream media, and it tends to respond to critics with derision or dismissal rather than collegial engagement. Evangelicals have been quick to criticize the process behind the Jesus Seminar and Darwinism. They have resisted the idea that complex scholarly issues could be decided by a majority vote among club members. Those same critical instincts need to be kept in place when evaluating claims of consensus on global warming.

• Compliance with the Kyoto Protocol "would reduce global warming by less than 0.2 degrees Fahrenheit by 2050 — an amount so tiny as to disappear in annual fluctuation and with no significant impact on consequences."

• Imposing an absolute cap on national or global CO_2 emissions in the absence of any low-cost abatement options would create substantial risks of job losses and economic disruption, whether or not permits are tradable.

• We still must determine how harmful CO_2 emissions are and, thus, the benefits of reducing them. But, as we have seen, many scientists, especially agriculturalists, believe that CO_2 should not be classed as a pollutant at all because of its benefits to plant growth.

• Most of the proposals for cap-and-trade now on the table would exempt most developing countries from the cap. Because large, rapidly developing countries like India and China are among the exempt, and firms in regulated countries could move operations to unregulated countries to avoid abatement or permit costs, the result would be to leave actual global emissions largely unaffected.

• Church leaders, evangelicals in particular, are concerned about climate change primarily because they fear its potential impact on the world's poor, especially in the tropics.... Put simply, poor countries need income growth, trade liberalization, and secure supplies of reliable, low-cost electricity.... We should emphasize policies — such as affordable and abundant energy — that will help the poor prosper, thus making them less susceptible to the vagaries of weather and other threats in the first place.

• The assumptions are that reducing carbon dioxide emissions would so curtail global warming as to significantly reduce its anticipated harmful effects (which we have just seen is false)....

• The world's poor are much better served by enhancing their wealth through economic development than by whatever minute reductions might be achieved in future global warming by reducing CO_2 emissions.

• The ECI's claim that "deadly impacts are being experienced now" is unsubstantiated.

• It is immoral and harmful to Earth's poorest citizens to deny them the benefits of abundant, reliable, affordable electricity and other forms of energy (for homes, cars, airplanes, and factories) merely because it is produced by using fossil fuels. Foreseeable forms of renewable energy (other than hydroelectric) won't provide *reliable, affordable* electricity at least for many years, in amounts that are adequate and necessary for modern hospitals, factories, homes, communities, and nations.

• Compulsory programs are not market-driven; they are driven by regulations, treaties, and rent seeking. But such programs appeal to politicians, who want to hide the tax and blame others for the soaring prices.... We believe the market is a better judge of cost effectiveness than bureaucrats and politicians.

• We know we have said this before, but it bears repeating: since energy is an essential component in all economic production, artificially restricting its consumption will drive down production, drive up prices, and reduce access to life-improving and live-saving technologies, harming the poor especially.

• Foreseeable global warming will have moderate and mixed (not only harmful but also helpful), not catastrophic, consequences for humanity — including the poor — and the rest of the world's inhabitants.

Beisner published another paper entitled "Global Warming: Why Evangelicals Should Not Be Alarmed" in 2007. Despite the growing amount of scientific evidence by this time, he continued to criticize even evangelicals who had expressed increasing concern.[1]

"In a documentary aired August 23, 2007, by CNN and titled 'God's Warriors,' Richard Cizik, vice president for governmental affairs of the National Association of Evangelicals, said about evangelicals who disagree with his urgent appeals for action to fight global warming: 'Historically, evangelicals have reasoned like this: Scientists believe in evolution. Scientists are telling us climate change is real. Therefore, I won't believe what scientists are saying. It's illogical. It's an erroneous kind of syllogism. But is that what's been occurring? Absolutely.'"

Beisner once again stressed his credentials, such as they are. "I (am) an evangelical professor of social ethics who has for 20 years specialized in the application of biblical world view and theology to environmental economics and written three books and edited a fourth in the field." He sums up his findings thusly:

> In short, all of these scientific developments — and many more — provide good reason at least to question, if not to reject outright, the popular claim that human action is driving catastrophic climate change.
>
> Bible readers should find these developments unsurprising. In at least three ways, Scripture has prepared us for them. First, in Genesis 8:21-22, God promised Himself never to allow the cycles that sustain human (and other) life on Earth to cease so long as the Earth remains. Second, in Psalm 109: 6-9 we read that God 'set a boundary' that the sea could not pass over. Third, fears of [human-induced global warming] suppose

---

1 E. Calvin Beisner, *"Global Warming: Why Evangelicals Should Not Be Alarmed,"* ecalvinbeisner. com, September 2007.

a fragile biosphere and land/ocean/atmosphere system that is inconsistent with these verses and with the Bible's teaching that a wise Creator designed the Earth to be a resilient, self-regulating system suitable for human habitation."

So, in the end, groups like the Cornwall Alliance are trying to bend science to fit in with their biblical views. These organizations receive moral support (and money) from equally zealous industries worshipping before the altar of free markets, unlimited growth that is becoming more and more unsustainable, globalization, deregulation, and — most importantly — windfall profits.

The George W. Bush years proved to be very good indeed for the very wealthy. The average tax rate paid by the richest 400 Americans fell by a third to 17.2 percent through the first six years of his administration, and their average income doubled to $263.3 million, according to new Internal Revenue Service data.

The 17.2 percent tax rate in 2006 was the lowest since the IRS began tracking the 400 largest taxpayers in 1992, although that group paid more tax on an inflation-adjusted basis than any year since 2000.

The drop from the 2001 tax rate of 22.9 percent derived largely from Bush's push while president to cut tax rates on most capital gains to 15 percent in 2003. Capital gains were 63 percent of the richest Americans' adjusted gross income in 2006, or a combined $66.1 billion, according to the data.

As an addict might say, we will soon find out that Denial is not a river in Egypt.

Chapter Seven — Climate Change, American History, and the Economic Meltdown

> "Modern man no longer knows how to foresee or to forestall. He will end up destroying the earth from which he and other living creatures draw their food."
> — Albert Schweitzer
>
> "I am pessimistic about the human race because it is too ingenious for its own good. Our approach to nature is to beat it into submission. We would stand a better chance of survival if we accommodated ourselves to this planet and viewed it appreciatively instead of skeptically and dictatorially."
> — E. B. White
>
> "Politics is not the art of the possible. It consists in choosing between the disastrous and the unpalatable."
> — John Kenneth Galbraith

The recent global and domestic financial crises have sent economists scurrying to review lessons that could be gleaned from the Great Depression. Surely this era would hold some key to unraveling our current economic troubles.

American history reveals that we would need to review our entire past as a nation to realize that we have been here many, many times before and consistently for the same reason. Large corporations have consistently acted badly when they could, bristled at any form of regulation that would affect immense profits, and were not shy about launching political intrigues and character assassinations of the worst kind in order to achieve their objectives.

Questions about and objections to the squandering of natural resources did not begin in recent decades or even with the pro-conservation stance of President Theodore Roosevelt. These issues have abounded over the past three centuries —

with little relief. Perhaps the major difference in the past was that a substantial number of the Americans were much more vocal and less quiescent in places like Kansas, for instance, than we are today. Few denied the obvious fact that the now forbidden term "class warfare" was an economic reality. And average Americans were not afraid to say so.

If the past is any indication, President Barack Obama will face a tsunami of criticism over any reforms and regulations aimed at addressing the climate change issue. His utopian vision that bipartisan approach to environmental problems will be successful has no precedence in American history. As his predecessor George W. Bush might have said, he has woefully "misunderestimated" the forces he is up against. In fact, all of his economic gurus have close ties with and have benefitted from the largesse of Wall Street.

American history in both the economic and conservation realms should now be apparent. American business concerns have and continue to exploit domestic and foreign resources at breakneck speed and banks extend easy credit during boom times. The economic system then either sputters (recession) or collapses (depression) in boom and bust cycles. The culprits have always been the same: greed and a lack of social conscience. The government (through its taxpayers) is then forced to rush to the rescue, forcing big business to conform to rational regulation. While the business community has no other options but bankruptcy, they kick and scream about their lack of market "freedom," even as the economic house is burning down and public funds are necessary to save them.

Throughout the history of American-style capitalism, large corporations have become masters at exerting political pressure and pushing through their agendas — or defeating issues viewed as onerous and expensive. Phase one involves election and party donations to assure "access" for arcane but profitable legislation as it travels under the radar. In the short term, politicians need money and, also in the short term, corporations must meet or exceed Wall Street expectations on a quarterly basis. In the long term — well, there is no long term. Should some threat arise that ultimately cannot be squelched, an expensive, finely-tuned public relations spin machine is unleashed. Often working through "think tanks" with benign puppy-dog names, the corporations attempt to shape public opinion through slight-of-hand or outright deceit. If the perceived foe cannot be vanquished by adverse media saturation (aided by the media's "he said/ she said" presentations), political canards that have proven effective for generations are resurrected. And if all else fails, the time comes for personal attacks and innuendo.

The well-orchestrated attacks by chemical companies and other corporate interests against biologist-turned-writer Rachel Carson in the 1960s provide a good example of the methods used by big business to dissuade, silence or discredit those who stand up to them. Unable to debunk the science and government reports which were checked a rechecked, the chemical industry was not above hinting that the unmarried Carson was probably a communist, was certainly an uppity and hysterical female, and could well be a lesbian. Carson's experiences after publication of her book *Silent Spring* are chronicled in *Rachel Carson: Witness for Nature* by Linda Lear.[1]

Any significant action toward climate change cannot be discussed without looking at the ramifications of the global economic crisis and the trillions being spent in the wake of the malfeasance of the American financial sector. Even President Obama's modest "cap-and-trade" proposal is now generally believed to be in jeopardy.[2] Ironically, harsh economic times will succeed in slowing down investment — public or private — for possible remedies to combat global warming on the grounds that they will be "bad for business" and, therefore, hinder job growth and economic recovery. The recession will be very good for those who deny we are on the brink of catastrophic manmade climate change and it lets off the hook industries that have resisted the costs of retooling of America. All of those additional corporate lobbyists might not have been needed after all. Conservative think tanks are fairly crowing that the average American no longer considers climate change a significant issue — the state of the economy, as former President Clinton correctly pointed out, trumps everything.

A simple yet complex question remains: What went wrong? As an astute historian might say, "Round up the usual suspects."

One thing is certain. The American public has had a difficult time comprehending how the economy flew off the rails. This is not the place for a full treatise on modern economics, but we'll digress for a bit to show how the economy can affect, or be used to affect, public opinion about other issues, and how politics — not the "democracy" we all believe in — are shaping this country.

Things began to take a turn in the wrong direction a decade ago with the passage of the Gramm–Leach–Bliley Financial Services Modernization Act near the end of President Clinton's tenure. Championed by then-Senator Phil Gramm of Texas and tucked into an appropriations bill where it could evade debate, the 385-page bill struck down the Glass-Steagall Act that had been enacted to protect the public in the aftermath of the Great Depression. Traditional banks had been complaining bitterly that the Glass-Steagall Act had denied them the opportunity to participate in a hot domestic market and handcuffed them in a competitive global marketplace. Now, they were set free. They turned into investment banks and were allowed to enter the lucrative derivatives market — with the additional advantage of little oversight. Derivatives are both a cushion and a gamble. But there was no clearinghouse holding collateral to settle a deal gone bad and no transparent records of who was trading what. Therefore, risk could not be assessed.

---

1 Linda Lear, *Rachel Carson: Witness for Nature* (New York: Henry Holt and Company, 1997.
2 Michael Barone, "*Obama Cap-and-Trade Will Meet Coal-Fired Energy Political Opposition,*" U.S. News & World Report, March 25, 2009.

"By appearing to provide a safety net, derivatives had the unintended effect of encouraging more risk-taking," according to Anthony Faiola, Ellen Nakashima and Jill Drew of the *Washington Post* in a story published on October 15, 2008.[1] The global derivatives market topped $530 trillion as of June 30, 2008. "When the housing bubble burst and mortgages went south, the consequences seeped through the entire web," the *Post* reported. "Some of those holding credit swaps wanted their money; some who owed didn't have enough money in reserve to pay."

According to the *Post* article, "Derivatives ... accelerated the recent collapses of the nation's venerable investment houses and magnified the panic that has since crippled the global financial system."

When the economy began to fall to its knees and unfathomable amounts of money were doled out to the financial sector (as well as industry), former Federal Reserve Chairman Alan Greenspan testified that even he did not fully understand the global financial markets.[2] Who was to blame? Barack Obama's economic advisor Lawrence Summers as well as Democratic icon and former Treasury Secretary Robert Rubin, President Clinton, and a bipartisan group of legislators had joined Greenspan in advocating the deregulation of markets and the dismantling of the firewall established by the Glass-Steagall Act.

Former presidential candidate Ralph Nader had often claimed that there was no substantial difference between the policies of Democratic and Republican elites. The story of the Financial Services Modernization Act would seem to support Nader's conclusions.

Senator Gramm was chairman of the Senate Banking Committee at the time. It becomes apparent that he sold the bill as something that would benefit American consumers by offering the convenience of a one-stop shopping center for all financial needs. Thrown in as a sop was increased consumer financial privacy. The real intent of the 385-page bill was rarely discussed and rarely questioned. The following statements were issued by Senator Gramm's office:

> The ... reason for doing this bill ... is to expand both the volume and the quality of financial services, and do it in a way that will end up producing lower prices for the American consumer....

> I think [the bill] represents the American legislative process at its best. It has resulted more from an effort to reach a logical conclusion than to satisfy various special interest groups.... My test is: What are we trying to do in the bill? Are we trying to benefit banks or insurance companies or securities companies, or are we trying to benefit consumers and workers?

---

1   Anthony Faiola, Ellen Nakashima and Jill Drew, *"What Went Wrong,"* Washington Post, Oct. 15, 2008.

2   *"Greenspan Admits 'Mistake' That Helped Crisis,"* Associated Press, Oct. 23, 2008.

The Senate approved the Gramm–Leach–Bliley Act by a vote of 90-8, doing away with Depression-era barriers separations between banking, insurance and securities. Gramm thanked his bipartisan banking committee, including Democrats Paul Sarbanes of Maryland, Christopher Dodd of Connecticut, John Kerry of Massachusetts, Jack Reed of Rhode Island, Chuck Schumer of New York, and John Edwards of North Carolina.

Posing as a great compromiser on an issue far removed from the issue of melding commercial banks with investment banks, Gramm also met repeatedly with Summers and Sperling to discuss the existing Community Reinvestment Act which provided low-interest loans to community groups. "I remain committed to a real, full-fledged sunshine amendment that will make public the CRA-related agreements between banks and community groups," he said. The meat of the bill — the repeal of the Glass-Steagall Act — was not mentioned.

President Clinton signed the Financial Modernization Bill on Nov. 12, 1999 with speeches all around. Obama's top economic adviser Lawrence Summers was among them. Gramm noted that when Glass–Steagall became law, the country believed that government was the answer, that stability and growth came from government keeping out of free markets. "Today," he said, "we are here to repeal Glass–Steagall because we have learned that government is not the answer. We have learned that freedom and competition are the answers. We have learned that we promote economic growth, and we promote stability, by having competition and freedom. I am proud to be here because this is an important bill. It is a deregulatory bill."

In preparation for introducing President Clinton, Summers effusively thanked Gramm and extolled the wisdom of their current decision:

> By turning a budget deficit that was threatening to exceed half a trillion dollars, ultimately, into a budget surplus, their steps and the steps taken by the Congress have freed up literally trillions of dollars of capital ... to be invested in American plant and equipment and in homes for American workers. But as important as assuring a large quantity of capital is assuring that capital is allocated efficiently and competitively throughout our economy. And that, of course, is the task of the financial system and is a task that the financial system will carry on more efficiently and more effectively to the benefit of Americans as a consequence of this legislation.

President Clinton then took the podium. After praising everyone involved, he said, "The future of our children will be very bright, indeed."[1]

Very bright indeed. Bright for some — Democratic and Republican movers and shakers and the banking and financial sectors in equal measure. For the average American, not so bright.

---

1   Ibid.

In February of 2009, Bill Clinton acknowledged, "I think that the only thing that our administration did or didn't do that we should have done is to try to set in motion some more formal regulations of the derivatives market. [But] they're wrong in saying that the elimination of the Glass-Steagall division between banks and investment banks contributed to this. Investment banks were already ... banks were already doing investment business and investment companies were already in the banking business. The bill I signed actually at least puts some standards there. And if you look at the evidence of the banks that have gotten into trouble, the ones that were the most directly involved in there ... in a diversified portfolio tended to do better.

"Some of the conservatives said that I was responsible because I enforced the Community Reinvestment Act, and they said that's what made all these subprime mortgages be issued. That's also false. The community banks, the people that loan their money in the community instead of buying these esoteric securities, they're doing quite well."[1]

A decade earlier Brooksley E. Born, head of the Commodity Futures Trading Commission from 1996 to 1999, had warned her peers on the President's Working Group on Financial Markets and called for tighter regulation of the financial system; they pushed back, as that was the very system that had earned them wealth and power.

Born was concerned about derivatives and she wanted to talk about regulation. Born testified before Congress at least 17 times that ignoring the risk of derivatives was dangerous.

"Greenspan, Rubin and Levitt were determined to derail her effort," the *Washington Post* reported, and Born was told to back off.

Washington Post staffers conducted over 60 interviews and collected transcripts of meetings, congressional testimony, and speeches in an attempt to answer the question: What went wrong? This tremendous debacle was truly a bipartisan effort among the Washington and Wall Street power elites.

"The future that Born envisioned turned out to be even riskier than she imagined," the *Post* continued. "The real estate boom and easy credit of the past decade gave birth to more complex securities and derivatives, this time linked to the inflated value of millions of homes bought by Americans ultimately unable to afford them. That created a new chain of risk, starting with the heavily indebted homebuyers and ending in a vast, unregulated web of contracts worldwide."

Ongoing attacks by Senator Richard Lugar, an Indiana Republican, Alan Greenspan, and others eventually left Born politically isolated. She left the Commodity Futures Trading Commission (CFTC). Once again, the rich and powerful had shot the messenger.

---

1 John Roberts, CNN interview with Bill Clinton regarding the Gramm-Leach-Bliley bill, Feb. 16, 2009.

The deregulation baton was then handed over to Gramm to guide through the legislative process. Summers, Rubin's successor at Treasury, still insisted on keeping the CFTC out of the swaps market. Gramm supported Summers, "holding out for stronger language that would bar both the CFTC and the SEC from meddling in the swaps market," according to the *Post*.

The act passed as a rider to an omnibus spending bill in December 2000, while the people and the press were still transfixed by the Supreme Court decision settling the presidential election in favor of George W. Bush just three days earlier.

As the *Post* commented, "...the act did not provide for any SEC oversight of investment bank holding companies. The momentum was all toward deregulation." It took only another five years for the dominos to begin to fall.

Clinton, Rubin, and Levitt (the SEC chairman from 1993 to 2001) have all made comments in the recent past to distance themselves from the derivatives market fiasco. Considering the bare-knuckle tactics used during the deregulation fight, some might find it difficult to chalk up trillions of dollars wasted, jobs lost, and lives ruined to a simple mistake.

Levitt joined the international powerhouse The Carlyle Group after leaving the SEC. In January 2001, he received the "Award for Distinguished Leadership in Global Capital Markets" from the Yale School of Management.

Levitt would have been considered the "top cop" overseeing financial markets. Yet in 2000, he seemed to be concentrating on "fair disclosure" issues of financial information between Wall Street analysts and corporations — in other words, monitoring talking-head analysts on television to see if they were on the payrolls of corporations.

Larry Summers, who succeeded Robert Rubin as Treasury Secretary from 1999 to 2001, is heralded in his government biography as having led "efforts to modernize the financial system" and insuring "the viability of the over-the-counter derivatives market."

Summers is now the top economic adviser to President Obama. According to a *New York Times* story written by Jeff Zeleny and published on April 4, 2009, the White House disclosed that Summers had earned more that $5 million in 2008 from the hedge fund D. E. Shaw and collected $2.7 million in speaking fees from Wall Street companies that received government bailout money.[1]

In 2008, Summers reported making 40 paid appearances, including a $135,000 speech to the investment firm Goldman Sachs, in addition to his earnings from the hedge fund, a sector the administration is trying to regulate. He also appeared before Citigroup, JP Morgan and the now-defunct Lehman Brothers.

"While Mr. Obama campaigned on a pledge to restrict lobbyists from working in the White House, a step intended to reduce any influence between the

---

1 Jeff Zeleny, *"Financial Industry Paid Millions to Obama Aide,"* The New York Times, April 3, 2009.

administration and corporations, the ban did not apply to former executives like Mr. Summers, who was not a registered lobbyist," according to the *New York Times.* "In 2006, he became a managing director of D. E. Shaw, a firm that manages about $30 billion in assets, making it one of the biggest hedge funds in the world."

The *Times* also reported that White House aide Michael Froman, deputy national security adviser for international economic affairs in the Obama Administration, received more than $7.4 million from Citigroup, where Froman previously worked.

Greenspan, as head of the Federal Reserve from 1987 to 2006, was in a sense the most powerful man in the world, with companies awaiting his abstruse oracles on the economy. Beginning in the Reagan era, he was a consistent champion of deregulation.

Greenspan also assured the public that the financial and real estate markets were sound, with most homeowners experiencing "a modest but persistent rise in home values that is perceived to be largely permanent." Somehow the $8 trillion home valuation wealth bubble slipped past him.

Not everyone has been as enamored by Greenspan's performance. Stephen Lendman of the Centre for Research and Globalization said in a 2008 interview that credit default swaps were "little more than casino-type gambling. Unregulated with no transparency in the shadow banking system that dwarfs the traditional one in size and risk."

"His successor [Ben] Bernanke did nothing to curb it," Lendman added.

"To a large extent, the U.S. crisis was actually made by the Fed," according to economist Jeffrey Sachs. The *Guardian* wrote in 2007 that globalization was "laying bare the contradictions of capitalism" and with its "unbridled economic activity" which was destroying "the world's climate, water supplies, farmland, forests, and fish stocks."

"Over the years, Mr. Greenspan helped enable an ambitious American experiment in letting market forces run free," commented law professor Frank Partnoy, adding that the Fed chairman "championed adjustable rate mortgages and ignored the clear fraud from the subprime ones."

The only player not interviewed in the *Post* investigation about the swaps debacle was Brooksley Born — the only person of vision who tried in vain to ring the alarm bells and ring them often. While the designation of "hero" is used too loosely in our society, Born would qualify for standing up to some of the most powerful people in America and, indeed, the world. Her efforts to save Wall Street from itself and to save the American people from real pain were repaid with scorn and public ridicule.

The Obama administration, which has pledge to overhaul the financial system, includes some former Treasury officials who opposed Born a decade ago. While Obama has often declared that he inherited the country's financial mess, his economic team includes members like Summers who played a significant role in deregulating the derivatives market and abetting that mess.

Economic supporters of American-style free market capitalism in Washington have long relied on rabidly free market economic "think tanks" such as the influential American Enterprise Institute for fiscal guidance. Peter J. Wallison, an Arthur F. Burns Fellow in Financial Policy Studies at AEI, specializes in financial market deregulation. A former lawyer in the Reagan Administration, gave a speech to the Federal Reserve Bank of Chicago (published on the AEI website in May of 2000) entitled "The Gramm–Leach–Bliley Act Eliminated the Separation of Banking and Commerce: How This Will Affect the Future of the Safety Net."[1] Let's see how accurate this prognosticator has been in the cold light of today.

> About 15 years ago, the debate over bank deregulation focused on the question: 'Are banks special?' A good deal of ink was spilled in debating this issue — with the opponents of a deregulation proposal advanced by the Treasury Department arguing that banks were special and shouldn't be allowed to affiliate with other financial activities. The principal opponents then — Fed Chairman Paul Volcker and New York Federal Reserve Bank President Gerald Corrigan — argued that banks were indeed special. Now the Fed supports the GLB Act, which is much the same in concept and form as the Treasury proposal that it opposed 18 years ago.
>
> By authorizing banks to affiliate with securities firms and insurance companies, Congress has in effect acknowledged that there is no *policy* reason to hold banks aloof from the economy at large. Although Congress tried to draw a line between finance and other kinds of commercial activities, this distinction cannot hold. Because of the weakness of its policy underpinnings, the GLB Act cannot be seen as anything more than an interim step on the way to complete elimination of restrictions on the kinds of organizations that can control or be affiliated with banks.
>
> Under these circumstances, there is little reason to retain a bank safety net. If banks are not special in the sense that they should be kept apart from and treated differently than other elements of the economy, it is questionable whether the government should take any more responsibility for their survival or prosperity than it takes for other financial institutions or business enterprises.
>
> ...[P]rivate structures, as has now been adequately demonstrated, can achieve payment finality without any government role.
>
> The Fed's role as a lender of last resort — at least insofar as it involves assistance to individual banks — is clearly a relic of an earlier time

---

1 Peter J. Wallison, *"The Gramm-Leach-Bliley Act Eliminated the Separation of Banking and Commerce: How This Will Affect the Future of the Safety Net,"* American Enterprise Institute for Public Policy Research, May 5, 2000.

and old theories about the vulnerability of banks.... The credit markets today are much wider and deeper than they were in the past, and well-regarded banks would have no difficulty obtaining short-term advances through the interbank market in the event that they suffer substantial withdrawals.

In the general case, if weak banks are unable to meet their obligations, they should be allowed to fail and investors to take their losses.

The key value the Fed contributes to stability of the current large dollar payment system is its guarantee of payment finality.... [O]f course, the Fed takes a significant risk. But is there a reason why the payment system must operate in this way? ...No..., a private payment system could be developed that would function equally well — perhaps better. The Clearing House Interbank Payment System, known as CHIPS, is an entirely private international interbank payment system that cleared over 57 million payment transactions, totaling almost $300 trillion, last year.... CHIPS is [a] demonstration that a wholly private large dollar payment system can function effectively and safely.... A similar private system in the United States could supplant the Federal Reserve.

And on and on.

If this analysis is correct, the safety net — as it was understood at the end of the 20th century — will wither away in the 21st. Along with it will go the policy reasons for intensive regulation of banks.

[W]hat could possibly be the policy reason for drawing a line between financial services and the rest of commerce. I contend that there is none, and mere fear of bigness is not a sufficient reason if financial conglomerates can get as big as they want.

With all due respect to Wallison, why didn't he save the American people a lot of money by simply offering up the CHIPS phone number and avoid a taxpayer bailout? A trillion here and a trillion there adds up and could perhaps have been put to better use — like saving the planet.

The Heritage Foundation, another free marketeering think tank, also exerted a great deal of influence during the George W. Bush years, endorsing the new legislation because it "represents the most significant deregulation of the financial services industry in over half a century," adding that "One estimate suggests that consumers would save $15 billion a year in fees levied on financial services thanks to greater competition and a more efficient financial service system."[1]

Rubin repeated this $15 billion figure at a hearing before the House Committee on Banking and Financial Services in February of 1997.

Not everyone was so enamored. In 1994, Travis Plunkett, the legislative director of the Consumer Federation of America, testified that the new bills "create

---

1 David C. John, *"Gramm-Leach-Bliley Act (S. 900): A Major Step Toward Financial Deregulation,"* The Heritage Foundation, Oct. 28, 1999.

concerns that this shadow banking system could put taxpayer-backed deposits at risk."[1]

A strange acronym began to appear in financial analyses that would soon take on greater significance — TBTF, or Too Big To Fail.[2] Wall Street was fully aware that, no matter what risks were taken, the government could do nothing else but ride to the rescue. The potential for profits was enormous while the risks were minimal — because Wall Street was rewarded for both success and failure.

Paul Krugman, an economist at Princeton University and recent Nobel Prize winner, recently set the Wayback Machine to the year 1982 and pointed his pen at The Great Deregulator — Republican patron saint Ronald Reagan. In a June 2, 2009, column published in the *New York Times*, Krugman asserted that Reagan was the guiding force behind the Garn-St. Germain Depository Institutions Act, a bill that "turned the modest-sized troubles of savings-and-loan institutions into an utter catastrophe."[3]

The bill deregulated the savings and loan industry and was a contributing factor to that industry's crisis in the late 1980s. It had broad support in Congress where it passed by a margin of 272-91 in the House of Representatives. It was Title 8 of the act that allowed adjustable rate mortgages.

President Reagan, of course, called the bill "pro-consumer" — a label that would later serve Senator Gramm so well. What was it that President Bush once said? "Fool me once, shame on — shame on you. Fool me — you can't get fooled again." While this wise old saying somehow became garbled with lyrics from a rock song by The Who, the actual quote, "Fool me once, shame on you. Fool me twice, shame on me," should come to mind. But at least Reagan was honest enough to call the bill "the first step in our administration's comprehensive program of deregulation."

If President Reagan would have had his way, *his* administration could have claimed credit for the Gramm–Leach–Bliley debacle.

"For the more one looks into the origins of the current disaster, the clearer it becomes that the key wrong turn — the turn that made the crisis inevitable — took place in the early 1980s, during the Reagan years," Krugman writes. "Attacks on Reaganomics usually focus on rising inequality and fiscal irresponsibility. Indeed, Reagan ushered in an era which a small minority grew vastly rich, while working families saw only meager gains. He also broke with long-standing rules of fiscal prudence."

---

1   U.S. Senate Banking, Housing and Urban Affairs Committee, "*An Examination of the Gramm-Leach-Bliley Act Five Years After Its Passage*," testimony of Travis Plunkett, legislative director of the Consumer Federation of America, July 13, 2004.

2   The TBTF acronym seems to have appeared about 1991 in The Golembe Reports by Carter Golembe in the article "Too-Big-To-Fail and All That." This seems to have replaced TLTF (Too-Large-To-Fail) cited in 1990 which just didn't seem to have the same ring to it.

3   Paul Krugman, "*Reagan Did It*," The New York Times, June 1, 2009.

Government debt skyrocketed under Reagan, but that wasn't the worst of it. Krugman observes, "The increase in public debt was, however, dwarfed by the rise in private debt, made possible by financial deregulation. The change in America's financial rules was Reagan's biggest legacy. And it's the gift that keeps on giving....[D]eregulation in effect gave the industry — whose deposits were federally insured — a license to gamble with taxpayers' money, at best, or simply to loot it, at worst.... By the time the government closed the books on the affair, taxpayers had lost $130 billion, back when that was a lot of money.

"But there was also a longer-term effect. Reagan-era legislative changes essentially ended New Deal restrictions that, in particular, limited the ability of families to buy homes without putting a significant amount of money down."

The explosion of debt made the United States vulnerable. There was too much risk and too little capital. Reagan and his advisers, Krugman asserts, forgot the lessons of the Great Depression "and condemned the rest of us to repeat it."

The New Deal. How the Republicans hated — and still hate — the New Deal. The Holy Grail lies in its destruction and sending Franklin Roosevelt's ghost to the ninth level of Dante's hell.

Economist and economic journalist Robert Kuttner testified before the House Financial Services Committee on October 2, 2007, saying that "Your predecessors on the Senate Banking Committee [in the 1930s] would be appalled at the parallels between the systemic risks of the 1920s and many of the modern practices that have been permitted to seep back in to our financial markets."[1]

The eight years of the George W. Bush Administration were bleak days for environmentalists but remarkably good days for deregulation advocates and corporate interests. Many sometime lobbyists seemed to filter in and out of government and business; lucrative jobs in the private sector awaited those in government who were friendly to business. MBNA Corporation, a leading proponent of banking deregulation, was the biggest donor to the Bush–Cheney ticket in the 2000 election cycle at over $3.1 million. MBNA was acquired by Bank of America in 2005 for $35 billion. While most of its cash was doled out to Republicans, over the years then-Senator and now Vice President Joe Biden received $214,100. AIG (American International Group) gave $1.37 million to Republicans in 2002 or 61 percent of their money; this shifted when AIG gave 68 percent to Democrats during the last election cycle, some $576,750.

The economic meltdown seems to have been a team effort, but the Bush Administration on its own did irreparable damage both domestically and globally. One example will demonstrate the general tone of his eight years.

---

1 Committee on Financial Services, U.S. House of Representatives, testimony of Robert Kuttner, Oct. 2, 2007.

In January of 2007, President Bush signed an executive order that gave the White House much greater control over the rules and government policy statements that protect public health, safety, the environment, civil rights, and privacy. Bush said that every agency must have a regulatory policy office run by a political appointee to supervise directives given to regulated industries. "The White House will thus have a gatekeeper in each agency to analyze the costs and benefits of new rules and to make sure the agencies carry out the president's priorities," according to Robert Pear of the *New York Times*. Pear said that the order "suggests that the administration still has ways to exert its power after the take-over of Congress by the Democrats."[1]

A major objective of the executive order, according to many observers, was to rein in the Environmental Protection Agency.

According to the *Times* story, "Consumer, labor, and environmental groups denounced the executive order, saying it gave to much control to the White House and would hinder agencies' efforts to protect the public." Peter L. Strauss, a professor at Columbia Law School, is quoted as saying that the order "achieves a major increase in White House control over domestic government.... Having lost control of Congress, the president is doing what he can to increase his control of the executive branch."

The executive order was an end-run around the laws enacted by Congress which typically gave agencies the power to issue regulations. Business groups hailed the initiative, according to the *Times*.

Wesley P. Warren, program director at the National Resources Defense Council, said, "The executive order is a backdoor attempt to prevent the EPA from being able to enforce environmental safeguards that keep cancer-causing chemicals and other pollutants out of the air and water."

According to the *New York Times* story, President Bush was maneuvering against criticism from environmental and consumer groups to name Susan E. Dudley (said to be hostile to government regulation) to be administrator of the Office of Information and Regulatory Affairs at the Office of Management and Budget. Some of Dudley's views are reflected in the executive order. In a primer on regulation written in 2005, while she was at the Mercatus Center of George Mason University in Northern Virginia, Ms. Dudley said that government regulation was generally not warranted 'in the absence of a significant market failure.'"[2]

The Mercatus (Latin for "market") Center is a conservative, zealously free-market think tank which became very influential during the Reagan and George W. Bush administrations. The chairwoman was Wendy Gramm, wife of Senator Phil Gramm. The center is especially devoted to environmental deregulation within the energy industry.

---

1 Robert Pear, "*Bush Directive Increases Sway on Regulation,*" The New York Times, Jan. 30, 2007.
2 Ibid.

On July 11, 2008, Dudley publicly objected to the EPA's analysis of various ways to control greenhouse gases under the Clean Air Act and transmitted the objections of four cabinet members and four other agency heads. All of these objections were published along with the EPA's analysis in response to an April 2007 Supreme Court ruling that the EPA had the authority to regulate greenhouse gases. For the administration to publish a document and disavow its own conclusions was an "extraordinary move" as the *Wall Street Journal* said.

"The National Academy of Sciences study of CAFE has many excellent elements, but in the end it fails to apply the fundamental economic principle of consumer choice. If consumers do not want to trade higher fuel economy for higher prices or fewer other features, government should not compel them to."

According to the *Mason Gazette*, staffers at Mercatus referred to rumored environmental regulations as "UFOs."

Wendy Gramm, like her husband, is a fierce advocate of the "free hand of the market" which has worked to both her husband's and her own financial advantage. Their devotion to deregulation has worked out very well for the financial sector but has wreaked havoc on the general working population and the environment. Wendy was chairwoman of the "regulatory studies program" at Mercatus. She was on Enron's board of directors before its collapse and is infamous for spearheading what is known as the "Enron loophole" so the company could avoid governmental oversight. Enron had been a financial backer of Mercatus. Previously, she held several positions in the Reagan Administration, including the Commodity Futures Trading Commission when it exempted from regulation Enron trading in energy derivatives. The "Enron loophole" was designed to exploit Enron's on-line commodities transaction system by freeing electronic energy trading from regulation by rescinding restrictions in place since 1922. Enron could then do as it pleased — which included bilking investors, destroying jobs and savings, and ruining lives.

Wendy Gramm had also previously been head of the Office of Management and Budget's (OMB) "Information and Regulatory Affairs." During 2002, the OMB drew up a "hit list" of existing federal environmental regulations it believed should be changed or rescinded. Of these, 44 had been suggested by the Mercatus Center.

Phil and Wendy Gramm have been amply rewarded by their paymasters and, according to the *New York Times*, have gone from humble beginnings to multimillionaire status. In 1995, while Phil Gramm was preparing his ultimately unsuccessful run for the Republican presidential nomination, he famously commented: "I have the most reliable friend you can have in American politics, and that's ready money."

As a senator, Gramm had developed powerful connections and, following his stint in government, flew into the waiting arms of UBS Warburg, the investment banking arm of Switzerland's largest bank, with the title of vice chairman. Gramm was a registered lobbyist for UBS from 2004 to 2008.

UBS has had its problems recently with the federal government. The bank had 20,000 well-heeled American clients with offshore assets alone worth some $20 billion. In 2004, UBS was fined $100 million by the Federal Reserve. The bank was hoping that Gramm's list of wealthy friends would add to the lucrative American client base seeking to open Swiss accounts in order to avoid paying U.S. taxes. UBS feted their new employee at a glitzy ceremony in New York, apparently rapturous over plucking a former influential American politician into their ranks. Gramm was a keynote speaker:

"The ability of men and women to move their money to protect it, their property and their freedom is one of the basic freedoms that people exercise on this planet," Gramm told the audience. He added that when it came to government poking its nose into private finances, "for those of us who love liberty, it is a nightmare...."

Gramm was also a major force in fighting Hillary Clinton's health insurance reforms in the 1990s.

The economic brain trusts that supported unregulated markets and actively worked on the Gramm–Leach–Bliley Act are largely still with us and have become influential members of President Obama's economic team during the current recession. The people who are now trying to fix the crisis are those who helped to bring the economy to this point in the first place. Little is made of this fact. The Democratic duo of Summers and Rubin joined Greenspan during the 1990s in a concerted push to deregulate markets. Gary Gensler, who worked with both Phil Gramm and Greenspan on this debacle, was nominated by Obama to head the CFTC. Robert Wolf, the CEO of UBS Americas, reportedly raised over $500,000 for Obama and bundled more than $370,850 for the candidate in 2008, making UBS Obama's fifth-largest donor, according to *SubmergingMarkets.com*. Obama also attended private dinners hosted by Wolf, according to the website.

Rubin was the head of the Treasury Department until handing over the reins to Summers in the late 1990s. Rubin already had a personal net worth of over $50 million in 1993 after working at Goldman Sachs. After his stint at Treasury, Rubin went to Citigroup where he collected another $115 million while acting as Citigroup director and senior counselor. These figures do not include stock options. Citigroup was rescued in November 2008 in a massive bailout by the U.S. government.

Rubin's close relationship with Greenspan was well known. In January 2009, Rubin was named by Marketwatch as one of the "10 most unethical people in business." After reviewing the facts, what stands out is that the recent economic downturn was exacerbated in a truly bipartisan way — with hands reaching across the aisle when it was to the benefit of politicians and financiers alike.

Obama frequently speaks of fostering the spirit of bipartisanship — and the Gramm–Leach–Bliley Act was a perfect example of mutual cooperation. Many of those involved — politicians, financial gurus, and titans of industry — made obscene amounts of money as political coffers were filled. The American public, however, did not fare so well. Joe Six-Pack is more likely to be trudging off to the unemployment office worried about his mortgage, the health care he can no longer afford, and the fight he just had with his wife over money. He can only sigh and curse his fate, blaming no one in particular. After all, Joe had played by the rules and those rules resulted in ruin. Joe doesn't realize that more than fate had been at work.

Treasury Secretary Timothy Geithner has also faced criticism for his close ties to the banking industry and the view that his bailout plans have been "overly generous" to the financial sector at taxpayer expense. His close personal ties to the financial sector date back to when Geithner was head of the New York Federal Reserve Bank. In an April 27, 2008, story in the *New York Times*, reporters Jo Becker and Gretchen Morgenson wrote that Geithner was hardly a Wall Street outsider who was "using an unprecedented amount of taxpayer money to try to save the nation's financiers from their own mistakes." [1]

Geithner "forged unusually close relationships with executives of Wall Street's giant financial institutions," according to the *Times*, adding that he was "often aligned with the industry's interests and desires." The story noted his particularly close relationship with Citigroup and Robert Rubin, his mentor who was then a senior Citigroup executive. In fact, Geithner had once been offered the top job at Citigroup.

According to the *Times*, Geithner's critics say that he "repeatedly missed or overlooked signs that [Citigroup] — along with the rest of the financial system — was falling apart" and he "failed to take actions on derivatives" although he tracked them during his tenure as president of the New York Federal Reserve. Critics argued that the New York Fed was "often more of a Wall Street mouthpiece than a cop."

"A revolving door has long connected Wall Street and the New York Federal Reserve," according to the *Times*, and that New York is crucial because it is the most powerful of the 12 regional banks that make up the Federal Reserve system.

Geithner's close personal relationship with Rubin has cast doubts on his objectivity as Citigroup, the world's largest bank, "dove deeper into mortgage-backed securities." In the social realm, Geithner played tennis with hedge fund managers and dined with the financial movers and shakers. Geithner told the *Times* these personal relationships did not influence his actions during his time

---

1 Jo Becker and Gretchen Morgenson, "*Geithner, as Member and Overseer, Forged Ties to Finance Club*," The New York Times, April 27, 2008.

at the New York Fed. Geithner said that he met often with Rubin but "I did not do supervision with Bob Rubin." As late as May, 2007, Geithner gave a speech to the Federal Reserve Bank of Atlanta and "praised the strength of the nation's top financial institutions, saying that innovations like derivatives 'had improved the capacity to measure and manage risk....'"

Citigroup alone would eventually receive $45 billion in direct taxpayer assistance. "[Geithner] also went to bat for Goldman Sachs, one of [AIG]'s biggest trading partners."

The world of Wall Street and the Treasury Department often intersect, with Treasury recruiting from Wall Street's ranks. Financiers and those who regulate them essentially belong to the same elite club and, over the years, have developed close personal ties.

This is a club that includes both high-powered Democrats and Republicans.

The economic downturn this club has engineered or allowed to happen will have real-life effects on our ability as a nation to fight climate change even marginally. Government bailouts have left unemployed and uninsured workers in no mood to bail out American industry as well, by paying for a transition to cleaner technologies. The cry will once again be for jobs at the expense of the environment. Industry will receive a free pass or a soft touch until we reach a climatic tipping point. By then, of course, it will be too late.

"Energy lobbyists ... note little public groundswell for climate action in the U.S. because of the economic crises," according to *ContrarianDirection.com*. "According to a recent Pew poll, global warming was dead last in a list of 20 national priorities."

According to Politico.com, the number of D.C. lobbyists now working to influence federal policy on climate change has increased in the past few years by 300 percent to 2,340 lobbyists — four climate change lobbyists for every member of Congress. The price tag for this lobbying effort is approximately $90 million.

"If democracy does not win this one, if the lobbyists win, perhaps the best we can do for our grandchildren is buy them a ticket to another planet," commented NASA climate expert James Hansen.

A recent PBS *News Hour* program featured a weekly discussion between Mark Shields and David Brooks who normally hash out their liberal/conservative differences. But on this show, the two men agreed on one issue: President Obama's "cap-and-trade" scheme to reduce greenhouse gas emissions — already viewed by most reputable scientists as inadequate — would become a victim of the troubled economy.

And in case anyone believes that intensive lobbying has diminished under the Obama Administration, this was dispelled by a recent study by the University of Kansas released in April of 2009.

"Even as President Obama vows to curb lobbyists' influence, the industry is booming as never before," according to an *Associated Press* article on figures compiled in the new report.[1]

Large companies that spent hundreds of millions lobbying successfully for the tax break enacted in 2004 received a 22,000% return on that investment — which the *Associated Press* story called "proof that for those who can afford it, hiring a lobbyist can pay handsome dividends."

"It clearly is a lucrative field for lobbyists," Stephen Mazza, who was involved in conducting the study, told the *AP*. "Congress wanted to create jobs, and what they probably did was create jobs for lobbyists."

Lobbying has seen "explosive growth" in recent years. Companies and special interest groups spent $3.42 billion lobbying Congress and the federal government in 2008, a 14 percent increase over the previous year.

Lobbyists lobby because lobbying works.

The data alarmed watchdog groups, which worry that ordinary Americans who can't afford representation by a well-paid lobbyist will lose out in debates with companies and interest groups who can.

Companies are "spending big money, but ... it pales in comparison to the potential profit they can reap if they're successful," said Sheila Krumholz of the Center for Responsive Politics, which tracks money in politics. She said she worries that those without lobbyists are being denied "a seat at the table."

The nonpartisan group also compared the amount spent by bailed-out banks on political contributions and lobbying with the amount of money they received from the Wall Street rescue fund, known as the Troubled Asset Relief Program.

American-style capitalism is sacrosanct in Washington and will remain that way, despite the rhetoric or even the good intentions of the Obama Administration. Mess with the money and you had better be prepared for a Washington-style smackdown.

President Obama, who once touted an energy policy that would save the world, has in a remarkably short time run up against an economic crisis and opposing political forces far more wily, experienced, and well-heeled than himself and the force of his personality.

In April of 2009, independent journalist Gwynne Dyer, who has written extensively on climate change issues, pointed out that Obama's new-found "pragmatic approach" would likely fall miserably short of substantial climate change policy.[2] Although he seems to fully understand the problems, the president-elect, he predicted, would become just another optimist caught up in the meat grinder of Washington politics and economic constraints — which have been further

---

1 Julie Hirschfeld Davis, "*The Influence Game: Firms got 22,000 percent return on lobbying cash in '04 tax change,*" Associated Press, April 9, 2009.

2 Gwynne Dyer, "Obama's Emission Cuts are 'Pragmatic Suicide,'" *The Toledo Blade*, April 7, 2009.

confounded by the financial shenanigans on Wall Street. The process, although expected, is painful to watch.

The new word in the Obama vocabulary, according to Dyer, is "pragmatic" — not what should be done but what is politically expedient.

"How will the Obama Administration reconcile political 'pragmatism' with scientific realities?" Dyer asks. He quotes Jonathan Pershing, the head of the U.S. delegation overseeing a new U.N. climate agreement. " 'There is a small window where they overlap. We hope to find it,' Pershing explained. But it doesn't really exist."

"A dozen wasted years later [following the 1997 Kyoto treaty], the climate problem has grown hugely, so this time everybody else is determined that the United States must be on board....

"But we recently learned what [Obama] thinks is 'pragmatic.' It is that the United States should cut its emissions back to the 1990 level by 2020."

Dyer cites a recent study released by the Hadley Climate Center in England, generally regarded as one of the world's most respected sources of climate predictions. The study shows that even rapid cuts in global greenhouse gas emissions, turning the 1 percent annual growth into a 3 percent annual decline within a few years, would still warm the planet by 3 degrees Fahrenheit by 2020.

"That is dangerously near the 3.6 degree rise in average global temperature that is the point of no return," Dyer writes. "Further warming would trigger natural processes that release vast quantities of greenhouse gases into the atmosphere from melting permafrost and warming oceans.

"These processes, once begun, are unstoppable, and could make the planet from 7 to 11 degrees hotter than the present by the end of the century.

"At those temperatures, much of the planet turns to desert, and the remaining farmland, mostly in the high latitudes, can support at best 10 percent or 20 percent of the world's current population. That is why the official policy of the European Union is never to exceed a couple of degrees of warming."

Dyer said the Obama Administration will fall far short of that goal.

"We are in deep trouble," Dyer concludes, "and 'pragmatism' will not save us."

For those who follow legitimate climate science and who fully understand the consequences of taking inadequate steps to effectively rein in our nefarious ways, the frustration and anger are palpable. The ignorance — if it is indeed ignorance and not simply pandering to the public or political payola — is nothing short of astounding.

In a remarkably blunt column appearing in the *New York Times* on June 28, 2009, Paul Krugman finally used a word that is completely apropos for the political deniers in the United States when it comes to combating climate change — "treason."[1] As the climate-change deniers never tire of pointing out, this Nobel

---

1 Paul Krugman, "Betraying the Planet," *The New York Times*, June 28, 2009.

Prize-winning economist and professor at Princeton University, this effete intellectual, is once again hammering the body politic with wacky protestations.

"So the House passed the Waxman–Markey climate-change bill," Krugman writes.

> In political terms, it was a remarkable achievement.

> But 212 representatives voted no. A handful of these no votes came from representatives who considered the bill too weak, but most rejected the bill because they rejected the whole notion that we have to do something about greenhouse gases.

> And as I watched the deniers make their arguments, I couldn't help thinking that I was watching a form of treason — treason against the planet.

> To fully appreciate the irresponsibility and immorality of climate-change denial, you need to know about the grim turn taken by the latest climate research.

Krugman is referring to a study released on May 20, 2009, by the MIT Center for Global Change Science.[1] The researchers conclude "that without rapid and massive action, the problem will be about twice as severe as previously estimated six years ago — and could be even worse than that."

The report states that the median temperature is expected to increase 5.2 degrees Celsius by 2100, with a 90 percent probability range of 3.5 to 7.4 degrees, compared to the median projected increase in a 2003 study of just 2.4 degrees, based on 400 runs of advanced computer modeling.

In fact, Ronald Prinn said that the modeling may actually understate the problem because of positive (and, in this case, positive means negative for the environment) feedback effects including melting permafrost in the Arctic and subsequent releases of large quantities of methane, a very potent greenhouse gas.

"The fact is that the planet is changing faster than even pessimists expected: Ice caps are shrinking, and arid zones spreading, at a terrifying rate," Krugman continues. "And according to a number of recent studies, catastrophe — a rise in temperature so large as to be almost unthinkable — can no longer be considered a mere possibility. It is, instead, the most likely outcome if we continue along our present course."[2]

The climate of New Hampshire could become that of North Carolina; Illinois becomes East Texas; and extreme heat begins to roll across the country with catastrophic results, stressing an already inadequate power grid and decreasing agricultural production — among other negative effects.

---

1 Ronald Prinn *et al.*, "New Analysis Shows Climate Change Odds Much Worse Than Thought," The MIT Center for Global Change Science, Massachusetts Institute of Technology, May 20, 2009.

2 Krugman, ibid.

At the June 26 House floor debate over the newly-named Clean Energy and Security Act, Rep. Paul Broun, a Georgia Republican and, remarkably, a member of the House Science & Technology Committee, received a round of applause from his GOP colleagues when he claimed that man-made global warming is a "hoax" with "no scientific consensus."

"Scientists all over this world say that the idea of human-induced global climate change is one of the greatest hoaxes perpetrated out of the scientific community," Brown stated. "It is a hoax. There is no scientific consensus." [1]

Brown also said that "[W]e need to be good stewards of our environment, but this is not it, it's a hoax" so, therefore, "This rule must be defeated."

Rep. Lloyd Doggett, a Democrat from Texas, who switched from opposing to supporting the bill, said he did so because he was tired of listening to members of "the Flat Earth Society" — Republicans who maintained that global warming was a theory and not a scientific fact — make "inane" remarks about climate change.

Krugman commented in his column: "I'd call this a crazy conspiracy theory, but doing so would be unfair to crazy conspiracy theorists."

He went on to write that opponents of the climate bill misrepresented the bill's economic impact "which all suggest that the cost will be relatively low."

"Still, is it fair to call climate denial a form of treason? Isn't it politics as usual?" Krugman asks. "Yes, it is — and that's why it's unforgivable. The existential threat from climate change is all too real.

"Yet the deniers are choosing, willfully, to ignore that threat, placing future generations of Americans in grave danger, simply because it's in their political interest to pretend that there's nothing to worry about. If that's not betrayal, I don't know what is." [2]

Mother Nature, as we are about to find out, rules with an iron fist mightier than the "invisible hand." We will do too little, too late, without the comforting excuse that we had not been warned.

Tick, tick, tick....

---

1 C-Span, U.S. House of Representatives floor debate over the Clean Energy and Security Act, June 26, 2009.

2 Krugman, ibid.

Bibliography

Big Dragon: China's Future: What It Means for Business, the Economy, and the Global Order, Daniel Burstein and Arne de Keijzer; 1998; Simon & Schuster.

Catastrophe: An Investigation Into the Origins of the Modern World, David Keys; 1999; Random House, Inc.

Climate Change: A Groundwork Guide, Shelley Tanaka; 2006; Anansi Press.

Collapse: How Societies Choose to Fail or Succeed, Jared Diamond; 2005; Viking Penguin.

Dry: Life Without Water, Ehsan Masood and Daniel Schaffer; 2006; Harvard University Press.

Endangered Species: How We Can Avoid Mass Destruction and Build A Lasting Peace, Stephen M. Younger; 2007; Harper Collins Publishers.

Every Drop For Sale: Our Desperate Battle Over Water in a World About to Run Out, Jeffrey Rothfeder; 2001; Penguin Putnam Inc.

Global Fever: How to Treat Climate Change, William H. Calvin; 2008; The University of Chicago Press.

Global Survival: The Challenge and Its Implications for Thinking and Acting, Ervin Laszlo and Peter Seidel, editors; 2006; SelectBooks, Inc.

How Much is Enough?: The Consumer Society and the Future of the Earth, Alan Durning; 1992; W.W. Norton & Company, Inc.

On the Brink: The Great Lakes in the 21st Century, Dave Dempsey; 2004; Michigan State University Press.

Rachel Carson: Witness for Nature, Linda Lear; 1997; Henry Holt and Company.

Running Out: How Global Shortages Change the Economic Paradigm, Pablo Rafael Gonzalez; 2006 (updated 2008); Algora Publishing.

Silent Spring, Rachel Carson; 1962; Houghton Mifflin Company.

Six Billion Plus: World Population in the Twenty-First Century, K. Bruce Newbold; 2007; Rowman & Littlefield Publishers, Inc.

The Bush Agenda: Invading the World One Economy at a Time, Antonia Juhasz; 2006; ReganBooks.

The Clash of Civilizations and the Remaking of World Order, Samuel P. Huntington; 1996; Simon & Schuster Inc.

The Discovery of Global Warming, Spencer R. Weart; 2003; Harvard University Press.

The Economy of China, Shu Shin Luh; 2006; Mason Crest Publishers.

The End of History and the Last Man, Francis Fukuyama; 1992; Avon Books, Inc.

The Growth of the American Republic, Samuel Eliot Morison, Henry Steele Commager and William E. Leuchtenburg; 1969; Vols. I and II, Sixth Edition, Oxford University Press.

The Road, Cormac McCarthy; 2006; Alfred A. Knopf.

The Two-Mile Time Machine: Ice Cores, Abrupt Climate Change, and Our Future, Richard B. Alley; 2000; Princeton University Press.

Twilight in the Desert: The Coming Saudi Oil Shock and the World Economy, Matthew R. Simmons; 2005; John Wiley & Sons, Inc.

Under A Green Sky: Global Warming, the Mass Extinctions of the Past and What They Can Tell Us About Our Future, Peter D. Ward; 2007; Collins.

Water: The Next Great Resource Battle, Laurence Pringle; 1982; MacMillan Publishing Co., Inc.

When the River's Run Dry: Water — The Defining Crisis of the Twenty-First Century, Fred Pearce; 2006; Beacon Press.